Concepts of Nonlethal Force

Concepts of Nonlethal Force

Understanding Force from Shouting to Shooting

Charles "Sid" Heal

Lantern Publishing & Media • Brooklyn

2020
Lantern Publishing & Media
128 Second Place
Brooklyn, NY 11231
www.lanternpm.org

Printed in the United States of America

Library of Congress Cataloging-in-Publication Data

Names: Heal, Sid, author.
Title: Concepts of nonlethal force : understanding force from shouting to shooting / Charles "Sid" Heal.
Description: Brooklyn : Lantern Publishing & Media, [2020] | Includes bibliographical references.
Identifiers: LCCN 2020022611 (print) | LCCN 2020022612 (ebook) | ISBN 9781590566350 (paperback) | ISBN 9781590566367 (ebook)
Subjects: LCSH: Nonlethal weapons. | Tactics. | Law enforcement.
Classification: LCC U795 .H43 2020 (print) | LCC U795 (ebook) | DDC 363.2028/4—dc23
LC record available at https://lccn.loc.gov/2020022611
LC ebook record available at https://lccn.loc.gov/2020022612

As with my other books, this one is dedicated to Linda, my wife of fifty years. From the girl I met on the way to Vietnam to our golden wedding anniversary, I could not have imagined how much of my future would depend on you. Through the years, I have learned that love is as strong as death and no one can find the meaning of life alone.

What God has joined together, let no one separate.—**Mark** 10:9

Contents

Introduction and Acknowledgments

As with any book of this sort, beginning with a common understanding of terms and definitions is important. In the case of nonlethal force, however, it is not entirely possible. Even the term "nonlethal" is controversial. Detractors are quick to point out, for instance, that people have died as a result of their use, so nonlethal is not accurate. Suggestions have included "less than lethal" or "less lethal." The problem is that neither of these is entirely accurate either, since "less than lethal" is just another way of saying "nonlethal" and "less lethal" implies less effective lethality.

In attempting to reconcile the discrepancies, some instructors have attempted to provide their own interpretations, but none has been widely accepted. Nor does it appear that there will be any consensus in the foreseeable future. The controversy doesn't just stop there though, because even the spelling is not universally accepted. Is *nonlethal* spelled as a single word or should it be hyphenated as *non-lethal*? Increasing the confusion still further is that the term nonlethal "weapons" is commonly used by the military but nonlethal "options" is preferred by law enforcement and they are not interchangeable, since all nonlethal weapons provide options but not all nonlethal options involve weapons. In fact, the law enforcement community nearly never refers to them as "nonlethal" and instead prefers either "less lethal" or "less than lethal."

Just when you think it couldn't get any more confusing, consider the advent of the TASER®, which is a trade name.[a] Reluctant to use a brand

a TASER® is an acronym coined by the inventor John "Jack" Cover, which he named after a childhood hero. Written at the turn of the twentieth century, the term "Thomas Swift's Electric Rifle" is the title of one of the more than a hundred volumes written by the Stratemeyer Syndicate. (The "A" was added to make a pronounceable acronym.) The modern device is currently manufactured by Axon (formerly TASER® International) in Scottsdale, Arizona.

to describe a nonlethal weapon, agencies have resorted to terms such as conducted electrical weapon (CEW), conducted energy device (CED), neuro-muscular incapacitation device (NMID), electro-shock weapon (ESW), electro-muscular disruption technology (EMDT), or stun gun. In fact, these terms are only representative and are by no means exhaustive.

These examples serve to highlight some of the controversy surrounding nonlethal force, which is the impetus for this book. Although it is impossible to bring closure to these issues, it is useful to at least provide a focus and reduce the distractions. To that end, I've been personally involved in this field for some decades now and have participated in discussions and workshops regarding many of the issues that have arisen in all facets of the use of nonlethal force. Accordingly, I've elected to use the following conventions.

First, I will write in the first person rather than the seemingly omniscient point of view required for more academic treatises. This will not only allow me to "speak" peer to peer but remove any doubt that any opinions are my own. Moreover, the merits of the work will stand or fall on how well I present the case and not on any esoteric dissertations, titles, or appendages. I will, however, cite my sources for follow-up and more in-depth research.

Second, I will do my best to simplify the issues for clarity, even to the point of generalization.[b] Many of the more contentious issues tend to rely on anecdotal incidents, hence generalization will not only provide a broader view but serve to emphasize the importance of context in the employment of these force options. Likewise, generalizations tend to be more useful in that their broader scope provides insight and direction for more situations without encumbering the reader with a myriad of details. It should go without saying that there will always be exceptions.

Third, whereas previous works have focused on major concerns such as acquisition, funding, and development, or conversely, on the individual weapons and their performance characteristics, this work is

---

b This will include the use of the male pronouns "he" and "his" for suspects. For any women who feel slighted that I have intentionally ignored their contributions to crime, please keep in mind that this is only for simplicity and clarity and no further inference should be taken. It should be noted that considerably more than nine of ten crooks and nearly all military combatants are male.

tightly focused on addressing those issues necessary for crafting policy and developing training, as well as identifying best practices for their employment. It employs a holistic overview rather than idiosyncratic case-by-case descriptions. The emphasis is on force options and weapons systems rather than specific types of weapons or munitions.

Fourth, I will try and make it readable for those "in the trenches." To this aim, I will use illustrations, anecdotes, and personal examples. Furthermore, even though it is impossible to completely isolate the various issues, to the extent possible, I will break them down into "bite-sized chunks." In fact, nothing is footnoted that is so common as to already be in Wikipedia. Only information intended to provide greater insight and meaning to the text will be footnoted. All other types of information, including citations and references, will be provided for students and the more serious reader as endnotes. It is hoped that this format will provide utility for both those who want fast and reliable information as well as those who seek an in-depth analysis.

Lastly, I need to say that this book is not for the faint of heart, easily offended, or bleeding hearts. I will try and be tactful unless clarity suffers. Sometimes, it is more important to be clear than tactful, however, and I will state upfront that *any* type of force is messy and ugly. Ignore what you "learned" by watching television. The formative years of my law enforcement career has been spent on the streets of South-Central Los Angeles. During that time, I have suffered a herniated disk in my back, a broken nose, a broken left hand, and a broken right ankle. I have been on crutches, braces, had three surgeries, and have no bursas in either elbow, not to mention so many stitches that I only think of them now when I see the scars.[c] Likewise, I have experienced the thrill and terror of following speeding ambulances and driving wounded deputies to the emergency room. There has already been much hyperbole, distortion, and equivocation on the subject of nonlethal force, and I do not wish to contribute to any further confusion, even at the cost of tact and diplomacy. Like all my partners, my experiences have made me a pragmatist. I have

c You know you've been in the ER too many times when the doctor prescribes a painkiller and you suggest an alternative.

no allegiance to funding, grants, brands, or esteem. All I ask from any nonlethal force option is that it works.

Sometimes, it is better to pose a question without an answer. Rhetorical questions have the advantage of causing a reader to think through the possibilities rather than providing obvious, but tedious and boring, answers. I will also admit to a bit of sarcasm now and then and repeating a silly question is one of the best ways of exposing the hidden agendas of the hypercritical and the nitpickers.[d]

The book is written for two distinct audiences, law enforcement and the military. Although they are both in the "force business" they have different perspectives, missions, and most of all, language and culture. Having spent a considerable portion of my life in both disciplines I will do my best to "translate" and explain concepts, issues, factors, and terminology that may be confusing to one or the other.

In order to "eat this elephant" one bite at a time, the book is broken into two parts. The first part is focused on issues. Each chapter is oriented to a single, ongoing issue relative to the development and/or employment of nonlethal weapons. Chapters are intended to provide enough information to inform the reader but without being exhaustive. Ancillary issues, discussions, and reference material are provided in the endnotes. This section is intended to assist in planning, crafting policy, and decision making. The second part is focused on the options themselves. Each chapter is focused on a separate category of nonlethal options and provides an overview of the various options available, and describes their characteristics, advantages, and shortcomings, as well as caveats and best practices.

In attempting to identify that "magic middle," I have continuously struggled with how much is enough and how much is too much? Because the field is so broad and complex, the emphasis for this book will be on the issues facing the user and those options most likely to be employed at the "street level." Likewise, a lot of attention will be on anti-personnel options because of their versatility, controversy, and difficulty. No one cares if you scratch a car with a running gear entanglement system but the same

---

d I'm not so crass or naïve to believe that everyone who disparages the use of nonlethal force is not genuinely concerned and sincere, but I've met a lot of knuckle draggers and boneheads, too.

"scratch" on a human with a net can result in stitches, slings, crutches, and broken bones. (Note: All first references to terms that appear in the Glossary of Terms are printed in bold, as above.)

To assist me in this ambitious project, I have received the advice and support of two close friends, RK Miller and John Stanley. Both are retired law enforcement officers and experts in nonlethal force, and both are authors in their own right. Most notably, both are instructors and practitioners with decades of experience. Nevertheless, they will be the first to admit that many of the complaints from detractors are valid. But what is the alternative? Faulting a method, or even prohibiting the use of a particular force option, may assuage a critic's sensitivities but has absolutely no effect on the problem in dispute. We can solve it with the best available nonlethal options, even with any inherent shortcomings, or revert to more primitive and predictable but commonly harsher measures. The standard is not perfection; it is the alternative![e]

Like my previous book, *Field Command*, the illustrations are provided by yet another street cop, Chris Branuelas. Chris is one of those rare of creatures who is both practical and artistic. As before, his suggestions and ability to portray esoteric concepts in a visual format is unsurpassed and provides far more clarity and understanding than my best attempts at verbal description.

This book is intended for police officers and the military. It is not a treatise on the rights of people to assemble to petition their government for redress of grievances; nor does it pose an opinion on any particular legislation (or lack thereof) that may bring people out onto the streets to protest. Although some of the implements discussed in this book have been employed by regimes over the centuries to oppress individuals or whole groups of people, it is not the intention of this book to advise or recommend to regimes that they use these weapons or tactics against innocent, law-abiding civilians.

As will be evident throughout this book, the contexts where nonlethal force is applied are overwhelmingly less dramatic and, frankly, more

---

e I often use this adage when teaching nonlethal force because it recognizes the fact that the true measurement of a force option is what would have been used without it.

complex and chaotic than those imagined by people who believe any force used by police officers always violates individual or collective rights.

We need to understand from the outset that there is no way to make force look pretty. Policing is not a genteel sport governed by umpires and referees; it is messy, confusing, and traumatic. This reality is why nonlethal force is invaluable when provocateurs hijack a peaceful assembly for their own political or criminal ends and put up barricades, smash windows, set cars on fire, or pick fights with officers, or one another. It is helpful when assailants—both singly and in groups—are high on drink, drugs, or adrenaline and do not respond rationally to requests or orders from law enforcement to vacate the area or disperse. A range of nonlethal items can prove vital in preventing wholesale rioting and looting. They offer resources for police officers and peacekeepers when groups of people are throwing Molotov cocktails, bricks, stones, and other items at them. They can stop joyriders, carjackers, pickpocket gangs, and other highly mobile offenders from fleeing a scene; and they can provide a less lethal means of preventing individuals who may be trying to get an officer to overreact or are behaving so erratically that they present a danger to themselves as well as to others.

The men and women of our police and armed forces need protection from those who would do them harm, and an ability to fulfill their duties without having to kill the people they are sworn to protect. This book is intended to help these officers make an informed choice on how to do this, while protecting citizens from being killed. In their efforts to reduce as much as possible bloodshed and loss of life, those who protect our public safety now have at their disposal an array of nonlethal force options. Enabling their appropriate, measured, judicious, prudent, and knowledgeable use brings us closer to a goal that is not only laudable in and of itself, but surely conducive to more trust between law enforcement and our communities.

As the author, I take full responsibility for what I've written. It is my best effort at explaining the issues and nuances of a complex and controversial subject. I have spent all my adult life involved in some facet of conflict resolution, to include being a combatant. Some will believe that this has tainted my perspective. Indeed, it has! It is what has made me that diehard pragmatist. I have little time and no allegiance for those who insist on a reality of their own making. I have personally experienced the failures in applying an unrealistic solution to a very real, and often dangerous, problem. To that end, this book is focused on providing practical, but humane force options in an inherently violent and amoral world.

While I accept the responsibilities associated with authorship, it would be disingenuous to accept all the credit. A more accurate description of my contribution is spokesman. The concepts presented here been extracted and polished from practitioners and experts from all over the globe. They have not been derived simply by study or inquiry from academia or government but are blood lessons from those with skin in the game. It was my original intention to recognize each of these by name, but as I compiled the list it was obvious that it would require its own appendix, which my publisher discouraged. I am left with recognizing their contributions and thanking them anonymously. Besides my unnamed colleagues in the USA, I must also acknowledge those from the United Kingdom, Canada, Brazil, Sweden, Switzerland, Israel, and Australia. I have done my utmost to be worthy of their confidence in expressing our development and understanding of these concepts. When history judges us, we want it said that we did the best with what we had, but wished for more.

Part I

The Reasons for Nonlethal Force Options

1

Nonlethal Confusion and Controversies

SO . . . WHICH IS it, "nonlethal" or "less lethal?" How about "less than lethal?" Or, would you prefer "low lethality," "soft kill," "mission kill," "minimal force," "low level force," "disabling effects," "anti-lethal," "low collateral damage," "controlled force," "compliance weapons," "nondeadly force," or maybe "limited damage"? Yes! At one time or another, every one of these terms has been used, and in many cases are still being used. In fact, a new term, "minimal force options," is garnering favor, but "nonlethal," remains the definition used by the military folks and "less lethal" or "less than lethal" are the most popular with law enforcement types.

Late in joining the fray but quick to recognize the need for common terminology as a basis for understanding, the US Department of Defense (DoD) directed that the term "non-lethal" be used.[1] Equally unsurprising is that it had no effect on the ongoing controversy in the law enforcement community, who persist to this day in endless debates as to the appropriate terminology. Currently, there is no altogether adequate description and certainly none that is universally accepted. Even more poignant, there appears to be no agreement in the foreseeable future. This is only the beginning of the controversies, however. To gain some understanding of the issues generating the many controversies surrounding nonlethal force it is worthwhile recognizing the roots of the confusion.

The Heart of the Problem

First, there is nothing inherently lethal or nonlethal about an inanimate object. Hence, there is no such thing as a nonlethal weapon. I will admit that this is immediately controversial, but it can be easily demonstrated. Some years ago, one of my former students related that he was on the witness stand regarding some issue with nonlethal force when the defense attorney wanted to make a point and asked him, "Is your sidearm lethal or nonlethal?" He replied, "It's whatever I want it to be." Besides sounding flippant, this was not the answer sought by defense counsel and he challenged the officer to explain. The officer then related that if he used his sidearm to fire a warning shot, it was no more dangerous than any other warning. Likewise, he could use it to shoot out a tire to prevent escape or attack. In these instances, it would be used as a nonlethal option. This strikes right to the heart of many of those who have not only oversimplified the issues but believe they are endowed with a hard-earned wisdom that actual practitioners, especially police officers, lack—the fact that they are unable to cite any better alternatives, notwithstanding.[2] Thus, the intent of the user, not the characteristics of the weapon, is the primary factor for whether a particular weapon should be considered lethal or not.[3]

Second, whereas a lethal weapon primarily attempts to defeat an adversary's ability to resist, a less lethal device attempts to defeat his will to resist. An adversary's ability to resist is visible, measurable, and concrete. For example, the military community uses algorithms to predict the degree of damage and destruction that can be expected from artillery rounds, bombs, and even small-arms fire. Every grenade, naval shell, rocket, artillery, or mortar round can be rated according to such things as the killing radius, wounding radius, and shrapnel radius. A person's will, on the other hand, is intangible. It defies measurement. When employing less lethal devices, abundant examples exist of people who have resisted despite being struck by different devices scores of times. Conversely, we've had crooks who surrendered after a nonlethal device was fired and missed! Likewise, "arcing" a TASER® or racking a shotgun has resulted in compliance with only the threat of force.[4]

Third, whereas a lethal weapon is judged on its effectiveness, a nonlethal weapon is judged on both effectiveness and safety. This is particularly problematic in that these are often competing objectives. Frequently, as effectiveness increases, safety decreases, and vice versa.

Indeed, most nonlethal options work by pain compliance; that is, inflicting sufficient pain to attain a desired change in behavior. The problem is that pain is highly subjective and what can be excruciating for one person may be only irritating to another. Moreover, given the relatively primitive nature of most nonlethal force options, pain is nearly always caused by some type of physiological distress, most commonly physical trauma. Since all uses of force are attempts at changing human behavior, coupled with the fact that the amount and type of force required is never completely apparent or universally applicable, some injury can always be expected.

Effective or Safe?

Just when you think it can't get more complicated, neither the terms "effective" nor "safe" have been satisfactorily defined in the force domain. For example, if you use a nonlethal weapon to knock a belligerent on his butt and he gets back up, a manufacturer will be quick to point out that the device performed as designed, but you won't find a street cop on the globe that will claim it was effective. The initial problem remains, and the likely next step is another application of force. Likewise, "safe" is a relative term. Safe from all harm? Safe in comparison to greater injury? Be assured, the fact that a bruised adversary survived an encounter that would have otherwise resulted in his death doesn't meet the standards of "safe" in many circles.

Another point of contention is the fact that nonlethal force can be used to make lethal force more effective. I have been confronted with this issue on several occasions, most notably by members of the International Committee of the Red Cross. As just one example, in November of 1999, I was speaking at the Jane's International Nonlethal Conference in London, England. The audience was composed of members of the military, law enforcement, academia, and political figures from throughout the world. The perspectives were as diverse as the audience, which included extremists.

During a question and answer period, I was asked about my role as the principal advisor for nonlethal force during the evacuation of UNISOM personnel from Mogadishu, Somalia several years previously.[5] I explained that the "**technicals**"[a] would drive noncombatants, especially women and children, in front of their formations to close with the Marines because of their reluctance to engage them for risk of injuring innocent people. A number of nonlethal force options were available to counter these efforts, among them the use of "**stingballs**," to disperse the noncombatants without serious injury, while leaving the technicals vulnerable. Aghast, one of the more prominent critics condemned this course of action because we were "targeting noncombatants," a violation of international law. When asked what she would have proposed, her response was a muddled and murky confusion that evaded the question without ever answering it. When other members of the audience challenged her answer, she reluctantly admitted to no better alternative.[b]

A Workable Definition for Nonlethal Weapons

Not surprisingly, there is also no universally accepted definition for nonlethal weapons. Discovering multiple definitions in use,[6] the DoD again decided to develop their own. Because of its clarity and simplicity, it remains my favorite and will be used throughout the remainder of this book:

> Weapons that are explicitly designed and weapons primarily employed so as to incapacitate personnel or materiel, while minimizing fatalities, permanent injury to personnel, and undesired damage to property and the environment.[7]

Whereas this definition brought closure to the dispute in the US military community, the controversy remains to this day in American law enforcement. One reason is that, while the US Department of Defense

---

a "Technicals" is a moniker for the Somali insurgents who used heavy weapons, such as .50 caliber machine guns or anti-tank guns, mounted on pickup trucks as a primitive form of mechanized infantry.

b This is also an example of technological advancements evolving faster than the laws and policies for their use. Failing to protect the noncombatants from certain death would put us in a position of being legally right and morally wrong.

provides guiding information and regulations for all the uniformed services and defense related industries, there is no such body in American law enforcement. Each region, jurisdiction, and agency remain free to use any terminology or definition it chooses. And they do! Furthermore, opinions remain strong and the controversy will probably never be completely put to rest. Even so, nearly all replicate the major themes in the DoD definition, especially the intent not to cause death or permanent injury.

One common difference in perspective is that much of the law enforcement community uses a definition that includes all aspects of nonlethal force and substitutes the term "options" for "weapons." This broader perspective includes any force alternative that can achieve tactical objectives without resorting to lethal force.[8] It also includes options that may not meet the conventional definition of a weapon. Water, for one example, can be used as an **anti-traction** technology in the form of ice, an **area denial** technology in cold weather as a chilling mist,[9] in a vaporized form to obscure vision, or as an impact **projectile** when propelled from water cannons. Even wetting down an area to make it muddy and unappealing to a lingering crowd can have a deterrent effect.[c] But, would water, per se, meet the typical understanding of a weapon? Even if it could, is there some advantage of labeling such benign materials with such an emotionally arousing term?

Likewise, **pheromones** can be used to attract noxious insects, such as mosquitoes, bees, or hornets, which in turn provides a nonlethal advantage as an area denial option. How far should the definition of a weapon be stretched and to what advantage? The use of the term "options" thus seems a reasonable and logical expansion of the nonlethal concept.

Exacerbating the confusion still further, nonlethal options are easier to define and describe than to build and employ. For just one example, there are currently no nonlethal options that incapacitate personnel. All commercially available nonlethal options debilitate, not incapacitate.

---

c This method was successfully used by Oakland (CA) Police Department during a demonstration and protest by Occupy Oakland in the fall of 2013.

They achieve their advantages by reducing the capabilities and/or will to resist, but they do not prevent a determined adversary or one addled by drugs or in a frenzy from persisting, which then requires additional applications and often harsher types of force. Thus, the definition exceeds the capability![d] What is easy in concept is difficult in application and a genuinely effective and truly safe nonlethal option remains a pipedream.

Why Isn't the Use of Lethal Force Diminishing?

One controversy that will not go away anytime soon is both valid and legitimate. A question often posed is, "If nonlethal force options are becoming better and more prevalent, why aren't incidents of lethal force decreasing?" The answer is quite simple but currently unprovable. Nonlethal options are still in their infancy and are quite primitive when compared to their lethal counterparts. The most common lethal force option is some type of firearm, which has more accuracy and greater range than any currently available nonlethal option. Understandably, a police officer confronting a deadly situation is more likely to prefer an option that is less likely to fail. Thus, they don't try and solve lethal situations with nonlethal force. The problem is that it is largely speculative because law enforcement agencies have not been required to report force incidents to any central authority and so there is little or no supporting evidence to compare or identify trends.[10]

One consolation is that while there is no common agreement of what to call nonlethal options or how to define them, there is nearly universal acceptance on the essence of their purpose. Every definition, in both the military and law enforcement communities and from every country with a nonlethal capability, conveys that the intent of nonlethal force is that it causes neither death nor serious injury.[11] Hence, the goal is clear, even if the means to achieve it are currently lacking. At least in this, we are all in agreement. When history judges us as a compassionate people with a reverence for human life, at least part of the conclusion will

---

d This refers to the anti-personnel role of nonlethal options. It is possible to incapacitate equipment, vehicles, and the like.

be based on our efforts at finding nonlethal ways of solving our perpetual conflicts.

Final But Not Settled

Coming full circle now, the question remains as to which term and definition to use. Once you understand the background you can "pick your poison," since none of them will be universally accepted anyway. For simplicity and clarity, I have chosen the non-hyphenated term "nonlethal" to be all-inclusive and the DoD definition for clarity and simplicity. For purposes of this text the terms and definition can be considered interchangeable with whatever you prefer. It is my hope, however, that you can now defend your choice with confidence.

2

A Brief History

OCCASIONALLY, EVEN POLITICIANS say something insightful. So it was that a presidential hopeful stated in a speech, “We can chart our future clearly and wisely only when we know the path which has led to the present.”[1] This is true for many things but especially so with nonlethal weapons. An understanding of what worked, what didn’t, and why is of enormous benefit in avoiding the pitfalls we’ve already struggled out of. In keeping with my attempt to be succinct, however, this will be only a synopsis of the critical path that has led to this point.

I need to begin with what I do *not* consider part of the historical context, because over the years I’ve read a number of works who have cited examples of ancient nonlethal weapons but that wouldn’t meet any modern definition. One of the oldest occurred during the late Bronze Age around 1446 B.C.E. when Joshua led the Israelites against the Canaanites at Jericho. Biblical accounts tell of the use of trumpets made from rams’ horns to bring down the walls of the city. Thus, a case can be made that a nonlethal acoustic device was used as an **anti-materiel** device.[2] Another oft-cited example is the “salting” of Carthage by the Romans after sacking the city and winning the Third Punic War in 146 B.C.E.[3] The intent was to make the ground unsuitable for agriculture and would thus work as a nonlethal area denial option. Still another example was the use of ground pepper to temporarily blind opposing troops by the Chinese.[4] Thus, it can be claimed that this constitutes a nonlethal option used as an anti-personnel device.

I would submit that none of these examples, nor any like them, are an application of nonlethal force because they were nonlethal *not* by intent but because of the technological shortcomings of the day or were used to make lethal force more effective and easier to apply. Accordingly, they would not meet any modern definition because they were not "explicitly designed and primarily employed with the *intent* of minimizing fatalities and permanent injury." That is not to say that history has no examples of nonlethal options being developed or used, only that they were so few and far between and had no lasting effect on changing the traditional focus from deadly force.

Early Examples

One of the most unequivocal examples of nonlethal force occurred about 178 C.E. when the Chinese employed lime dust as a control agent to quell a peasant riot. Using horse-drawn chariots equipped with bellows, fine limestone dust was dispersed into the wind.[5] Likewise, a case can be made that the first nine of the ten Egyptian plagues described in the book of Exodus in the Holy Bible were nonlethal options. Nevertheless, there can be little doubt that, until very modern times, killing an adversary has been the preferred method of imposing one's will.[6]

The early use of truncheons by London police would certainly be an example of nonlethal weapons that would meet the modern definitions and can be reliably dated back to the second decade of the nineteenth century and probably earlier than that. Whether referred to as truncheons, billy clubs, nightsticks, or batons, they remain in use throughout the world today in a wide variety of materials and configurations, arguably as the most common nonlethal option available to police anywhere in the world. They are also the most primitive, and only the skill and intent of the person wielding them prevents serious injury or death. The fact that they remain popular speaks loudly for the primitive nature of other alternatives that would displace them, given a commensurate degree of effectiveness and cost.

Lacking a bellwether event, it is tough to pin down the first time another nonlethal option meeting any modern definition was successfully employed, but one strong candidate occurred in 1912 when French police successfully used **tear gas** to capture a gang of Parisian bank robbers and also as a **riot control agent**.[7] The US Army's Chemical Warfare Division conducted extensive research on **chloroacetophenone (CN)**[a] and promoted it for civilian use. Consequently, the concept of using it in law enforcement applications proved appealing and, by the end of the 1920s, police departments in Chicago, San Francisco, New York, and Philadelphia were purchasing supplies of tear gas.[8] CN remained the favorite agent by American law enforcement throughout the early 1960s, when it was widely replaced by **2–chlorobenzalmalononitrile (CS)**. Unlike CN, which takes its common name as an abbreviation of the chemical, CS takes its name from the US chemists who created it in 1928, Ben Corson and Roger Stoughton. CS has the dual advantage of being both safer and more effective than CN and quickly became the preferred chemical by the mid-1970s. Currently, there are about fifteen chemical agents designed for controlling riots, but CS remains the favorite and is in wide use today.

Despite the success with tear gas, history records nearly no further advances in nonlethal force options for more than thirty years.[9] This is in stark contrast in an age of significant advances in other areas, such as the invention of the laser, LEDs (Light Emitting Diodes), supercomputers, handheld calculators, ATMs, robots, satellites, and the lunar landing.

By the 1960s, police in the United States were confronting large civil disturbances resulting from those advocating for their civil rights and those protesting the Vietnam War. Likewise, the British police and military were experiencing similar situations in Hong Kong with pro-Communist activists and sympathizers. The need for effective nonlethal options was becoming increasingly apparent and more difficult to ignore. It was reported that the British used short wooden dowels fired from shotguns to ricochet into the legs of rioters as early as 1958.[10] Not unexpectedly, severe injuries are not only possible from these so-called "**knee knockers**," but likely. Even so, as primitive as they were, they were less injurious than lethal alternatives and are manufactured and

---

a CN was invented in 1871 and so was fairly well known at the time. Sometimes called a Riot Control Agent (RCA), CN is only one of several of the most common agents collectively referred to as tear gas. The term is used generically because one of the common effects of these agents is lachrymation (teary eyes).

sold today. When confronted with the "troubles" in Northern Ireland, the British continued to develop nonlethal projectiles. Substituting rubber and plastic for the original wooden dowels, they developed projectiles capable of being fired from larger caliber launchers, most notably the 37mm size. One of the earliest was made of hard rubber and first used in Northern Ireland in 1970. By the end of 1974, an estimated 55,000[11] of these so-called **rubber bullets**[b] had been fired.[12]

Through the 1970s, both the US and UK continued developing nonlethal weapons with limited success. Other countries were also conducting their own research but were largely focused on using nonlethal options to augment lethal force. One impressive success was the use of a **flashbang**[c] by the Israelis during the rescue of their hostages at Entebbe, Uganda on July 4, 1976. Although early uses were focused on increasing the effectiveness of lethal force, they were soon adopted by law enforcement agencies throughout the world and remain the only nonlethal option capable of supporting dynamic entries when arresting barricaded suspects and rescuing hostages. On some occasions, they have also been used in controlling mobs during riots.

The next milestone was the advent of the use of **pepper spray** as an irritant. The active ingredient, oleoresin capsaicin, is found in all types of peppers[d] and is better known as **OC** or just pepper spray. OC has the advantage of being quicker reacting than either CN or CS and works equally well on animals (except birds) and humans. Consequently, it was first used by delivery personnel and post office employees as an animal repellent. When used as an anti-personnel spray, it also has the advantage of less cross-contamination.[13] Whereas pepper spray became available to American law enforcement in the early 1970s, it wasn't widely adopted until the FBI authorized its use by their special agents in 1987, after

b Modern nonlethal projectiles are now made of rubber, plastic, wood, **foam**, and a variety of other materials, but the term *rubber bullet* has become both a generic term and buzzword for any nonlethal projectile.

c A flashbang is a pyrotechnic device which, when ignited, emits a loud bang, bright flash, and pressure wave. It is also referred to as a flash/sound diversion, **distraction device**, or **diversionary device**.

d That used for nonlethal purposes is derived almost entirely from Cayenne Peppers, and occasionally Habanero Peppers.

which it was quickly embraced by nearly every law enforcement agency in the United States.[14] Arguably, it is the most widely accepted nonlethal force option adopted by domestic law enforcement since the baton in the 1820s.[15]

A relatively new compound, called **PAVA**, is beginning to compete with OC. PAVA[16] is an organic compound but is synthesized rather than harvested from plants. Like OC, the active ingredient is an analogue of capsaicin, but it enjoys the advantage of better quality control and is more heat-stable than OC. PAVA is already in wide use in the UK and increasing in popularity in the US. Although OC and PAVA are very popular as a spray against belligerents, CS remains the preferred agent for riot control and barricaded suspects because it is a better area contaminant and requires less time for decontamination.

The Sea Change

Even as late as the early 1990s, nonlethal weapons were limited in both capability and use for both the military and law enforcement communities. Those that were available and would eventually prove valuable were neither well known nor widely accepted. This situation changed dramatically, however, when two events gained worldwide attention. The first was the 1993 siege in Waco, Texas of the Branch Davidian compound by American law enforcement. The operation lasted fifty-one days and resulted in the deaths of seventy-six people, including twenty-one children. Janet Reno, the US Attorney General at the time, asked both the Pentagon and the CIA to help the US Department of Justice find better nonlethal technologies.[17]

Contrasting the failure of law enforcement was the success of the US Marines' use of nonlethal options in resolving a military situation that historically had required lethal force, when it evacuated the UNISOM personnel from Mogadishu, Somalia in the spring of 1995.[e] For months afterwards, the world press was enthralled that an organization that had long prided itself as being the most deadly on a battlefield had been the

e For more information, see Chapter 3, "So! Why Bother?"

first to incorporate nonlethal weapons in all aspects of planning and executing their mission. Suddenly languishing nonlethal programs were reinvigorated, and efforts were taken to organize disparate programs to bring focus.[18]

About this same time a new "extended-range **impact munition**," more often called a **stun bag** or **bean bag**,[19] provided American law enforcement a quick and inexpensive alternative to nightsticks and pepper spray. Although substantially more injurious than chemical sprays, they provided a margin of safety because they could be employed against combatants at longer ranges. They quickly gained popularity in the American law enforcement community[f] and remain one of the principal nonlethal weapons in their arsenals.

Just prior to the twenty-first century, one more nonlethal option gained great popularity when a new TASER®[20] was introduced in 1999. TASERs® had already been in limited use for nearly thirty years but with mixed results. Some may remember that before Rodney King[21] was beaten with batons, he shrugged off the effects of a TASER®. The new TASER®, however, was not only more ergonomic and easier to employ, it had much more power than older models. In fact, it remains the only commercially available nonlethal option that even approaches a level that might be considered incapacitation. Departments who began experimenting with it reported seemingly incredible success stories and it gained immediate and widespread acceptance in the law enforcement community. Within a decade it was not only being used by law enforcement in the United States, Canada, and the United Kingdom, but in more than forty other countries.[22]

One of the problems with developing new nonlethal technologies in the United States is that most of the funding for new projects, especially the more exotic, is from the Department of Defense. Notwithstanding notable successes, The DoD is encumbered by restrictive international accords and treaties. Consequently, some potentially useful technologies, especially those involving **chemicals**, are unable to be fully explored.

---

f They are often referred to as the bread and butter of law enforcement nonlethal force options.

Law enforcement, on the other hand, experiences situations every day that would benefit from better nonlethal options and without the regulations imposed at a national level, however, they are hampered by comparatively meager funding.

While improvements continue to advance in nonlethal capabilities, the two decades of the twenty-first century have provided virtually no new breakthroughs that have progressed beyond prototypes and pilot projects. By and large, improvements have been limited to enhancing existing technologies rather than creating new ones. Furthermore, none of the most promising nonlethal technologies on the horizon appear likely to enjoy the widespread success enjoyed by stun bags, pepper sprays, and TASERS®. The interest remains high, however, and with the increasing dissatisfaction of the shortcomings with those currently in use, the advent of more effective and safer nonlethal weapons is inevitable.

3

So! Why Bother?

WHAT WITH ALL the complexity and confusion, the more pragmatic among us are asking, why bother with nonlethal weapons at all? To be sure, humankind has been solving these same kinds of problems for thousands of years without them. Nevertheless, when dealing with some situations, especially belligerent individuals and aggressive mobs, the ability to achieve a successful outcome cannot be attained by mere force. If it were that simple, more force would automatically equate with a greater likelihood of victory. What is more likely to lead to success in law enforcement and peacekeeping operations is not the amount of force but rather the type of force[1] and how it is perceived. Without an ability to employ nonlethal force, the answer is defaulted to lethal options. As the American psychologist, Abraham Maslow noted, "If the only tool you have is a hammer, you tend to think of every problem as a nail."[2]

History reveals at least two seminal events, more than 175 years apart, that still strongly influence the development and employment of nonlethal force in resolving contemporary conflicts.

The Peterloo Massacre

This watershed event occurred two centuries ago on a hot and cloudless summer day near Manchester, England. The Napoleonic wars had concluded just four years earlier with the Battle of Waterloo, ending more than a decade of war. England was experiencing an economic depression with chronic unemployment and high food prices resulting in a series of political rallies to demonstrate discontent and demand reforms in Parliament. On Monday, August 16, 1819, a large crowd, estimated between 60,000 and 80,000 people, many of them women and children, gathered on St. Peter's Field. The field was in the process of being cleared of brush and trees to enable a section of the street to be constructed. Piles of brushwood were stacked at one end of the field.

The sheer size of the crowd, who admittedly were noisy and boisterous, unnerved the local authorities, who called upon military authorities to disperse it. (No effective police force existed at the time.) With drawn sabers, the cavalry charged the crowd. The first casualty was a two-year-old boy being carried by his mother when she was struck by one of the mounted troopers. One of the men killed was a veteran who had fought at Waterloo. All in all, an estimated fifteen people were killed with as many as 700 injured, at least 160 of them women.

Although the crowd had been dispersed from the field within about ten minutes, riots immediately erupted in the streets of Manchester and surrounding communities, which continued through the next day and resulted in still more deaths. Journalists who were present published accounts throughout the nation, resulting in widespread revulsion and condemnation. The debacle was labeled the Peterloo Massacre.

This event is recognized as transformational in how riots would be forever after controlled and is often cited as the catalyst for modern policing, especially with regards to crowds, mobs, and riots. For our purposes, however, there can be no doubt that had the use of force been the only criterion necessary to determine success, the outcome was decidedly in favor of the government. The authorities succeeded in accomplishing their tactical ends but failed, catastrophically, in any strategic sense. No longer would overwhelming force be the sole criterion for determining success.

Operation United Shield

In the early 1990s, a massive famine in Somalia, in the Horn of Africa, was greatly exacerbated when gangs seized food and blocked the distribution of relief supplies and aid from other countries. Despite concerns of being drawn into an unremitting conflict, US forces began an airlift of relief supplies to avoid mass starvation. Even this, however, proved too little and UN ground troops, including the United States, were eventually required to prevent thefts by the heavily armed local marauders. Although the operation was initially successful, the lack of even a semblance of a unified government stoked fears that the situation would revert as soon as the troops left.

In June of 1993, one of the factions[3] killed twenty-four Pakistani troops and in October of that year, US troops were also engaged, resulting in eighteen American deaths and seventy-three wounded, as well as the loss of two Black Hawk helicopters.[4] With a civil war that had been underway for more than a decade and no end in sight, the UN Security Council decided unanimously to withdraw.[5] It was in this context that Operation United Shield was forged to withdraw the last UNISOM personnel from Mogadishu in the spring of 1995. Commanded by United States Lt. Gen. Anthony Zinni, USMC, and assisted by a coalition of other countries,[6] the amphibious task force landed in Mogadishu, Somalia on March 1, 1995.

Recognizing the high probability of a bloody conflict and the strong likelihood of civilian casualties, Lt. Gen. Zinni had opted for a novel employment of nonlethal force to separate predatory combatants from civilians and encourage compliance without the necessity of resorting to lethal force. Marines spent six weeks of intensive training on-board ship and were equipped with an array of nonlethal options, to include batons, pepper spray, stingballs, flashbangs, **sponge grenades**, as well as more exotic lasers and foams. More importantly, the use of nonlethal force options was incorporated into the planning, preparation, and implementation at all echelons and phases of the operation.

Despite several lethal engagements, the results of the operation were an unqualified success and the Marines, who, as stated in Chapter 2, prided themselves on being the most lethal military force on any battlefield, were justifiably pleased in sparing civilian casualties. Nor did the success go unnoticed as media outlets throughout the world extolled the virtues of nonlethal force.

Operation United Shield is recognized as the bellwether event that energized contemporary efforts in resolving conflicts without being compelled to employ deadly force. A year later, the US Department of Defense issued a policy for nonlethal weapons[7] and a year after that the

Joint Non-Lethal Weapons Directorate (JNLWD) was established. The significance was also recognized by the law enforcement community, and a surge of interest and new technologies became available and were quickly adopted, continuing throughout the rest of the decade.

Advantages of Nonlethal Options

Contrary to the belief of some, a law enforcement or military agency that has a nonlethal capability gains advantages over one that does not. When force is viewed and reviewed, as is always the case nowadays, even a failure in the use of a nonlethal option can be a success in that it sends an implicit message of restraint. It is virtually impossible for faultfinders and critics to make a credible case for a rash or impetuous act, or even a lack of patience or compassion. The fact that the technology proved too primitive to provide the same effectiveness as lethal force is not the fault of the user. Instead, it implicitly conveys restraint and a reluctance to use harsher methods, even though they are more effective. And so, the mere attempt makes the effort noble. In all situations where force is required, the pathway to the moral high ground is secured with nonlethal options.

The case for nonlethal force options is generally oriented around six advantages over lethal options:

- The first is that they are more humane. Although this may seem simplistic, it is, after all, difficult to make a case for a humanitarian effort while killing the people you are sent to protect. Whereas members of the military will experience this during peacekeeping operations,[a] and similar operations other than war, law enforcement officers experience this on a daily basis.

- A second advantage is that nonlethal force options allow a commander to exert more control over a situation. Because

a I realize that peacekeeping operations are only one of the many diverse operations other than war with which the military is routinely involved. For you law enforcement types, there are also operations focused on peacemaking, peace enforcement, protection and security, recovery, noncombatant evacuations, and so forth. For clarity and simplicity, however, let me make the point quickly and with a minimum of verbiage.

nonlethal options require substantially less provocation before engagement, a commander can provide a quicker response and intervene at earlier and less dangerous stages of an escalating situation. To some extent then, a situation can be influenced to promote a more favorable outcome without the brutality of lethal force.

- A third advantage is that nonlethal force options provide a commander with much more flexibility and freedom of action. With them, a commander need not accept the terms of engagement solely as presented by an adversary but can shape the situation.[8] No longer constrained to apply lethal force and "repeat as necessary," a commander can tailor a response to more appropriately fit the circumstances.

- A fourth advantage is that the employment of a nonlethal force option is far less likely to provoke others. Bystanders, for example, are less likely to be sympathetic toward people who defy a law enforcement officer or peacekeeping force but are not killed. Furthermore, should it become necessary to resort to lethal force, the fact that nonlethal options had proven ineffective not only supports a need for escalation, but provides an implicit, and almost irrefutable, message of restraint.

- A fifth advantage is similar in that nonlethal force options are also less likely to raise public outcry. Let's face it, the imposition of *any* force is controversial and public support is often key in resolving a difficulty without escalating it even further. Even Napoleon acknowledged that "public opinion is the ruler of the world."[9]

- Finally, the skillful use of nonlethal options can force an adversary to declare intentions. This is extremely valuable in that the most difficult problem in using force, in either law enforcement or peacekeeping operations, is not how much or what type to be used, but rather whether it should have been used at all! The confusion and ambiguity inherent in these situations create an

> uncertainty that will always be questioned later. Because this uncertainty is always accompanied with risk, an officer or soldier may be scorned for acting too soon or too harshly, or conversely, killed for acting too late or too mildly.
>
> For example, consider a situation with a potential adversary approaching a checkpoint and ignoring orders to stop. The question arises, is he continuing because he needs to get close enough for the weapon he is concealing to be more effective? Or, does he simply not understand the verbal commands to stop? Without some type of nonlethal deterrence, either way this scenario results in some innocent person being killed. When nonlethal weapons are available, however, a sentry or law enforcement officer can knock this guy on his butt with an impact munition. If he gets up and runs away, he remains alive. If he gets up and continues, we can reasonably infer hostile intent and respond accordingly.[10]

Despite their shortcomings, nonlethal options will continue to grow in popularity. As a result, not having them available, or at least seriously considering them, will diminish credibility for those responsible when using force in asserting humanitarian compassion and respect. This has already been noted in law enforcement circles during civil court proceedings after someone was killed because nonlethal alternatives were not available, especially when a neighboring jurisdiction has them. A conspicuous lack of a nonlethal capability sends a tacit but distinct message of indifference or negligence. As they say, it is the same tide that raises all ships and we will be judged by what others have done or are capable of doing. To carry the metaphor one step further but well into the military environment, the ripple started by Lt. Gen. Zinni in 1995 has resulted in the swell of interest in nonlethal options that we are experiencing today. The use of nonlethal force options declares a dignity and reverence for human life and provides moral options in inherently amoral environments.

4

Using and Justifying Force

ALTHOUGH THE DISPUTE over what to call nonlethal options has been intense, it is by no means the only controversy surrounding them, nor even the most contentious. Fundamental to employing nonlethal alternatives is a thorough understanding of a concept called the **force continuum**.[1] Historically, military objectives have been achieved by killing or destroying an enemy. Force was always deadly; hence effectiveness was judged only to the extent and speed at which death or destruction could be introduced.[2] A huge gap existed between presenting a threat and carrying it out.[3] The law enforcement community had long been using a tool called the force continuum to explain the concept and application of force to lay juries and the military quickly adopted it for similar reasons, albeit their primary audience was a concerned populace rather than a jury.

For our purposes, force is best understood as the exercise of strength, energy, or power in order to impose one's will.[4] It can be focused on a person or a thing, as in using force to push a car. In resolving a conflict, however, it is always focused on one or more people. Without question, force is the most scrutinized aspect of any conflict and frequently a point of contention. It behooves everyone, but especially planners and decision makers, to thoroughly understand force as a concept, in all its forms and uses.

The Force Continuum

When force is viewed as a continuum, an array of options present themselves. That also means that the issues become more complicated. The first complication is that force can be both increased and decreased.

This is not very practical when lethal force is the only alternative, but the recent prevalence of nonlethal force options not only made it practical but nearly impossible to disregard. Like a rheostat switch, force could now be dialed to a desired intensity and then reduced as necessary and then increased again. This new ability caused anxiety in some circles since the binary options of either/or for lethal force now had a nearly infinite number of variations when including nonlethal options. Furthermore, since adversaries remained alive and complaining, the appropriateness of force could be judged long before an objective was achieved. Accordingly, not only did the application of force become more complicated, public opinion now weighed heavily concerning everything from the suitability of weapons to how they were wielded, against whom, under what circumstances, for how long, and so on. This was a problem long understood by the law enforcement community, where complaints of unreasonable force or excessive force[a] are not uncommon, but relatively new in military environments.

Today, more so than at any other time in history, it is not just how force is used but how it is perceived that is important. The military quickly dubbed the term the "CNN effect."[b] A third complication arose as nonlethal options became more abundant, because they are extremely diverse. Besides the more well-known projectiles,[c] they take on stranger forms, such as nets, aerosols, slimes, foams, smokes, liquids, odors, sounds, lights, and even **radio frequency energy**. How do you compare such differences? How do you incorporate them into a plan?

The role of attorneys not only took on increased significance but was greatly expanded. Moreover, the input of public information officers

---

a As members of law enforcement will be quick to attest, there is a difference between the terms *unreasonable force* and *excessive force*. Unreasonable force is use of any force when it is unjustified. Excessive force is that which is required but deemed to be more severe than is necessary in either kind or duration.

b At risk of being simplistic, the CNN effect is simply a theory that holds that the power of the press in forming public opinion can be ignored only at great risk, so much so, that adverse accounts (true or not) can not only be influential but decisive in nature.

c These have been generically, and inaccurately, described in the media as rubber bullets. In reality, they may be rubber, plastic, wood, foam, silica, and composites of many other materials.

was needed. This was another complication that the law enforcement community was more accustomed to than the military. Lastly, judging the effectiveness of nonlethal weapons is far more complicated than the lethal variety. Whereas lethal weapons are judged effective when they cross a threshold of incapacitation, nonlethal options are specifically designed and primarily employed not to ever cross that threshold and so remain in limbo between ineffectual and deadly.[5]

The utility of a continuum for understanding and depicting force conceptually, let alone explaining it to a lay audience, provides comprehension and clarity lacking with other means.[d] The beginning of the force continuum, sometimes referred to as the force spectrum, is initiated by a threat, whereas deadly force takes its proper position at the other end. Nonlethal alternatives allow a commander to increase and decrease the amount of force necessary to accomplish a mission. Movement up and down the force continuum is generally continuous and seamless, yet a careful examination reveals some broad categories.

Entry into the force spectrum generally begins with a threat of some sort. While a case might be made that a threat is not force, the stronger argument is that, without force, threats lack credibility, and so lose their influence. It should be apparent then, that without credibility there can be no threat.

Threats come in two types: an "expressed threat," such as when a commander makes known the consequences of defiance; or an "implied threat," in which the nature of the consequences is left to the imagination of an antagonist. Of the two, the implied threat is far more powerful. Although there are several reasons for this, the most predominant is because what we can do and what we are willing to do are usually farther apart than an adversary realizes. Arguably, *anything* that you do that requires your adversary to contemplate the consequences of his actions becomes a force multiplier in its own right. Even our mere presence creates an escalation of force because it creates a condition that requires our

---

d Although other methods, such as wheels and matrices, are often substituted for the continuum, especially in the law enforcement community, they are all derivatives from the concept of a continuum. In addition, regardless of how they are depicted, they form a continuum of force options from mild to harsh.

adversary to contemplate his actions. Thus, an implied threat is implicit in every encounter. This condition prevails throughout the spectrum and should be exploited to the maximum extent possible.[6]

The next major step along the continuum involves physical force of some type, but which is not necessarily coercive in nature. Generally, this includes devices that engage an antagonist strictly on his own volition without specific intervention by a member of a peacekeeping force or law enforcement officer. Examples may include concertina, **caltrops**, barbed wire, **sticky foam**, or aqueous foam enhanced with oleoresin capsicum or covering caltrops, barbed wire, or other like obstacles. They are typically emplaced before an engagement in attempts to "harden" a vulnerable target.[e] They are placed relatively low on the force continuum, not because of the amount of injury likely to be sustained, but because they are benign without the willful defiance of the individual attempting to thwart them.

Further on the continuum would come munitions that cause physical discomfort but fall short of inflicting serious trauma. Examples of these options would include flashbangs, tear gas, pepper spray, and the like. This category also includes body strikes from fists, elbows, knees, and feet. Although the discomfort or injury may be substantially less than that suffered from a caltrop or concertina wire, the employment of these options requires a decision to intervene and they are thus subject to the idiosyncrasies of the individuals employing them. Factors such as training, maturity, discipline, prejudice, fatigue, fear, and judgment all play a part in their application and require them to be viewed more closely than those options that involve only one will.

Still further on the continuum are those munitions that inflict trauma with a potential for severe injuries. Examples include batons, **saps**, stingballs, bean bags, batons, foam and **pellet** munitions, and so

---

e Hardening a target is a concept better understood by the military and is often part of a "**shaping operation**." These nonlethal devices can be employed before an anticipated engagement as part of a larger plan. A shaping operation can be defined as any series of actions taken in anticipation of an engagement or tactical operation designed to promote accomplishment of strategic objectives. They enhance success by negating or mitigating potentially adverse effects while strengthening or increasing potentially favorable factors.

forth. They are generally the point on the force continuum that separates nonlethal from deadly force.

Furthest along the continuum are lethal options. Although the particular conditions that merit deadly force should be identified, lethal and nonlethal options should always be regarded as part of the force continuum and not as separate options altogether. This avoids ambiguity and confusion as to when they are authorized. Many situations rapidly escalate from less dangerous circumstances before requiring deadly force to resolve. An individual who is free to employ a variety of options is more likely to be proactive, retain the initiative, and be quicker to recognize situations requiring deadly force than one compelled to examine a situation isolated by either/or parameters.

To be sure, this rationale for placing options into a force spectrum may seem simplistic, but it is as good as any I've seen and better than most.[7] One problem occurs when people believe that the force spectrum is linear, meaning that a harsher measure should be preceded by a milder one. This is a recipe for disaster and the main reason that law enforcement agencies have moved to other means of describing and prioritizing force options. The situations in which nonlethal options will provide an advantage are always somewhat unique and unpredictable. Even more critical, they are temporary. Failing to use force decisively and promptly can easily result in escalation, and so choosing a force option is not amenable to rigid policies and algorithms. The US Supreme Court has recognized the difficulties and established that the force used need not be the least amount, only that it was reasonable under the circumstances.[8] Thus, depending on the circumstances, shooting a suspect armed with a knife may be just as reasonable as using a TASER®. This is another issue where the concept is easier than the application.

Effects-based vs. Behavior-based Perspectives

There is one other issue worth mentioning, however, and that is that where you choose to place a particular force option on a spectrum depends on whether you believe the predominant criterion should be the amount of injury likely to be sustained by an adversary or the amount of defiance anticipated.[f] These are commonly referred to as "**effects-based**" and "**behavior-based**" rationales. The effects-based model uses the anticipated injuries resulting from the nonlethal option being considered, whereas the behavior-based model is focused on the threats and actions that will justify a particular level of force. More simply, effects-based rationales use the amount of injury likely to be incurred as the basis for prioritizing the different force options, while behavior-based rationales use the degree of defiance of the adversary. These two rationales can justify the placement of a force option, such as a TASER®, very low on the force continuum in an effects-based model, or very high in a behavior-based model. Remember that the lower in the continuum, the more likely a particular option is to be used.

The effects-based rationale predominates by far in US law enforcement, with an estimated 80 percent or more of the agencies adopting this model. From my experience, however, behavior-based models are much easier to justify, both in the criminal and civil courts, as well as the court of public opinion. In fact, this is the position most often expounded by civil libertarians.

In recent times, some agencies have gathered options into groups and allowed a practitioner to choose from any in the group rather than a single option. This allows more latitude by the practitioner in the uncertainty of the unfolding situation. This method is applied nearly exclusively to a behavior-based rationale and depicted as a matrix or a wheel. Nevertheless, it remains a continuum since force is both explained and depicted from mild to harsh and the selection of an appropriate force option is based upon both provocation and risk. In crafting a workable force continuum, it is important to understand and be able to explain your rationale regardless of which philosophy you agree with.

Another justifiable concern arises with accusations of overuse of nonlethal force. This occurs because the rules of engagement for nonlethal options seldom require the same level of provocation before justifying their

---

f Although I realize that some will take exception to "anticipation," and insist on words like "exhibited" or "displayed," finding out what is intended is one of the advantages of a nonlethal option. Likewise, waiting for an incontrovertible exhibition of intent is not only accompanied with significant risk, it cedes the initiative to the adversary.

use as their lethal counterparts and so are more likely to be used. Human nature being what it is makes this a valid concern. People naturally try to find the easiest and safest way to do something and the use of force is no exception. As a veteran of many street fights, I can personally attest that even the winner can have broken bones and stitches and that avoidance becomes more and more appealing with age and experience. Notwithstanding, it is not only highly subjective but controversial. Understandably, there is no angst when the alternatives were harsher but becomes more difficult to excuse when milder options were ignored. Both the military and law enforcement communities handle these types of issues with a combination of training, supervision, and policy.

5

Injuries from Nonlethal Force Options

THE PRINCIPLE GOAL of nonlethal options is to prevent death or serious injury. In reality, however, deaths are possible, and some type of injury can be expected. It would seem prudent, then, to gain an understanding of issues like what types of injuries are probable, how likely they are to occur, and what can be done to minimize them. It is, of course, impossible in a single chapter or even a single book to list every potential injury, much less how to avoid them. Of necessity then, this chapter will focus on identifying what causes the most common injuries from nonlethal options, to whom and under what circumstances they occur, as well as those proven measures for avoiding them.

Orders of Effects

One of the most important considerations in understanding injuries from nonlethal force is the nexus to the application of the force itself. These are identified as **first-**, **second-**, or **third-order effects**. First-order effects are those caused directly by the application of the force. For example, an application of OC spray will cause redness and topical pain in the eyes and contaminated skin, whereas getting hit with a baton will result in localized pain and bruising. Second-order effects are those that occur from the use of force indirectly, such as from falling. Although the root cause is the application of the force, the proximate cause is the circumstances and environment in which the force was applied.

From a user's perspective, first-order effects are unavoidable. There is nothing a user can do to change how a force option works. For example, if a chemical agent causes shortness of breath and/or pain and redness of the skin, there is nothing that a user can do to eliminate them. They are a function of the force option itself and need to be anticipated and planned for; but the nature of the effects is integral to the application of that particular force option. Accordingly, all first-order effects are the responsibility of the developer. Although a user can exert some control—with factors like range and aiming point for impact munitions, or **exposure**, duration, and concentration for chemical agents—only by changing the nature and characteristics of the force option can first-order effects be alleviated.

Second-order effects are contextual; that is, they are dependent upon the unique set of circumstances and environmental factors present when a nonlethal force option is employed. Second-order effects are entirely the responsibility of the user. Without exception, the user is responsible for those consequences that can be reasonably anticipated from an application of force. Thus, using a nonlethal option that impairs sight or balance on an adversary who is in danger of falling or one that impairs movement of an individual who is in danger of drowning, creates an expectation that the user is aware of the potential adverse consequences.[1] As might be expected, second-order effects can be far more severe than those directly caused by any particular nonlethal force option. This is especially so with those with milder effects like CS and OC sprays or TASERS®, which cause only minor injuries as first-order effects but can result in fatalities with second-order effects.

Third-order effects are those that exacerbate an existing health condition. These effects are neither directly related to the force nor dependent upon the circumstances or environmental factors, but rather inherent in the unique health conditions of the subject. A number of serious injuries and deaths have resulted indirectly from nonlethal force when a previous health condition was exacerbated but blamed on the nonlethal force option employed. These health conditions may be illnesses, such as diseases of the heart and lungs, or intoxication, as with stimulants, such as methamphetamine or cocaine. These have been most prevalent with the use of chemical agents, especially pepper spray, and the use of electrical devices, such as TASERS®. Understandably, a person with asthma who inhales pepper spray is more likely to experience more severe effects than a healthy person.

Likewise, a person who has amped his heart rate to 200 beats a minute through the use of an illegal substance and is subjected to *any* kind of force, can expect an inordinately adverse reaction. Notwithstanding, some are quick to attribute any increased injury or death directly to the imposition of nonlethal force, especially the specific force option, while ignoring any individual vulnerability or responsibility. Accordingly, third-order effects are the responsibility of the individual. There can be no realistic expectation of detecting or compensating for medical conditions or intoxication in an incident serious enough to require the use of any kind of force.

Double-edged Swords

The use of nonlethal force options is not without risk. In fact, one of the criteria for a selection of a force option is the likelihood of injury to the user because some are more likely to result in injuries than others. This is

TYPE OF INJURY	CAUSE	RESPONSIBILITY	INJURY REDUCTION
1st Order Effect	Directly by the application of force	Developer	Change the method or nature of the force option
2nd Order Effect	Contextual circumstances	User	Anticipate and compensate for adverse effects
3rd Order Effect	Exacerbation of existing health condition	Subject	Avoid situations that increase the risk of injury

Figure 1: Injury Orders of Effect

With the single exception of threats, nonlethal force options always result in some type of injury. These can be reduced but never eliminated and an understanding of the nature of these injuries is important. First-order effects are injuries resulting directly by the application of force. These are the responsibility of the developer. When a TASER® leaves puncture wounds, for example, only the developer can devise an alternative. Second-order effects are injuries that result from the context in which the force is used. A subject on a roof or balcony, for example, may fall from being struck with an impact munition. While the first-order effects inherent in the use of force will result in an injury, the second- order effects from falling can be far more serious and even fatal. The most controversial injuries result from third-order effects. These are pre-existing conditions that make a subject particularly vulnerable to a force option. Like a diabetic baker or anaphylactic beekeeper, only the subject can take precautions to avoid the related injuries.

not a result of some accident or malfunction but rather because nonlethal force options require the user to close with the adversary. A baton and an impact munition, for one good example, function identically; that is, they use impact to create pain. The engagement range for a baton,[a] however, is arm's length, whereas that for an impact munition can be extended to well beyond fifty feet. Even though they function in the same manner, the impact munition is far more "**forgiving**" for a user if it is not completely effective.[2] The term most often used to describe this factor is "**standoff**." With no exceptions, nonlethal options that have greater standoff capabilities are safer for users than those that require closer ranges.

Another factor resulting in injuries to users is related to the duration of a confrontation and the number of "doses"[b] of force required to gain compliance. Currently, there are no force options, including lethal force, which have proven successful every time. Many confrontations require repeated applications of force. The likelihood of both suspects and users being injured when a nonlethal force option has not proven immediately effective rises dramatically with both the duration of the event and the number of failures. It would seem sensible then, that when force is necessary, it should be applied decisively rather than escalating from milder methods. Somewhat ironically, this results in less injury to both suspects and users.[3]

Where More Than *What*

Not surprisingly, most injuries from nonlethal force options result from those that use impact to function and are nearly as likely to result from first- as second-order effects.[4] Although the injuries are usually just bruises and mild abrasions, serious lacerations and punctures are also possible. Research and experience have revealed that more than the range or the type of projectile, *where* a person is hit has a greater effect on the severity of injury than *what*

a Or truncheon, billy club, nightstick, sap, and any other description or configuration.

b Like defiance or effectiveness, nonlethal force is nearly impossible to accurately measure. As used here, a dose refers to a single application of a measured amount of force, regardless of whether it is from being struck with an impact device, exposed to a chemical agent, or incurred from an electrical impulse.

was used to strike him. For example, of the deaths and severe injuries that have been reported from impact munitions, nearly all have been from strikes in the chest region, either by a projectile penetrating the chest cavity or by breaking a rib that lacerates a vital organ.[5] One of the best solutions has simply been to change the aiming point from "center mass" to the belt buckle.[6] Whereas penetrations of the abdominal cavity are still possible, life-threatening injuries are greatly reduced by avoiding organs like the heart, lungs, and spleen. Furthermore, when injuries do occur (and they will), the urgency for immediate life saving measures is greatly reduced and so victims are able to be medivacked and treated at hospitals.

Impact munitions that fire multiple projectiles, such as pellets, or multiple baton rounds preclude any ability to choose which part of a body to avoid. Because of their ability to simultaneously strike numerous adversaries, impact munitions are appealing for handling mobs and riots and injuries to the face and eyes of people located far behind an adversary are possible. When these munitions are necessary, one of the safest and most effective methods has been to skip-fire the munitions directly in front of an adversary. It is safer because the greater saturation of projectiles near the adversary results in most of the pellets or batons striking the intended target and prevents them from continuing down range. It is more effective because more energy is transferred to the adversary as he is hit with more of the projectiles.[7]

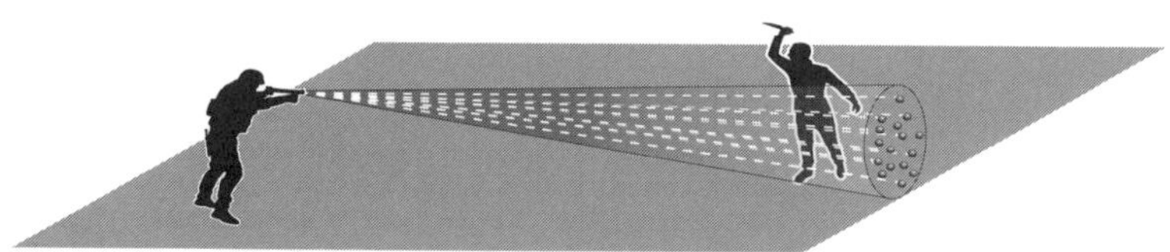

Figure 2: Beaten Zone

Some launched impact munitions propel multiple projectiles. A "stinger" round, for one example, fires hard rubber pellets as the nonlethal equivalent of buckshot. The pattern resulting from these multiple pellets is often referred to as the "beaten zone." The term is taken from their lethal counterparts and will vary with the nature of force option and the range. More rarely, the term can also refer to the ballistic deviations from single projectiles launched multiple times. Direct fired munitions can be problematic because, inevitably, some of the projectiles miss the intended target and continue in flight striking unintended targets.

The effects of all chemical agents currently being used as nonlethal options are relatively mild and leave the affected person capable of caring for his own safety, although admittedly impaired. Consequently, the injuries associated with nonlethal chemical agents are nearly all second-order effects.

OC, CS, and even CN are relatively short-lasting and dosage dependent. Individuals that have been exposed to an agent in a **plume cloud**[8] are far less contaminated than those who have been directly sprayed. By far, more people are hurt because they have tripped, fallen, or been run over by other people attempting to escape the effects than from the actual chemical agent. This has long been known in law enforcement and when chemical agents are used to quell riots or suppress the actions of a mob, avenues of escape for members of the mob are calculated into the deployment plan.

CS is the preferred agent for these purposes because it is persistent only when people remain in the contaminated area. After moving to a clean air source, the effects last only about fifteen to twenty minutes and require no additional decontamination procedures. In fact, one of the best decontamination measures has proven to be simply facing a stiff wind.

The effects from OC are both faster acting than CS and more severe. So, too, is the decontamination. Decontamination of OC can take forty-five minutes or longer, which is why OC is seldom used for contaminating large areas. OC is also very heavy and difficult to keep aloft, making it a poor area contaminant.

These attributes are desirable against single belligerents, however, and nearly all American law enforcement agencies rely heavily on OC sprays for individual confrontations. Even under ideal conditions, most sprays have a range of fifteen feet or less,[9] requiring close proximity to a belligerent to be effectively applied. Moreover, OC spray only works when it gets in the eyes, generally necessitating that the adversary be facing the user. The duty experts here are law enforcement officers who commonly use a distraction of some sort when applying pepper spray lest the adversary take effective countermeasures, usually as simple as turning their head or blocking the spray with their hands.

Once applied, however, some of the more belligerent combatants become enraged and will immediately attack, and a lot of the injuries to all parties occur during this period. Taking evasive action then becomes a sound consideration. Conversely, suspects who have anaesthetized themselves with drugs or alcohol may exhibit little or no discomfort.[10] As a general rule, if OC is applied correctly but didn't work the first time, it won't work at all, and a different force option needs to be quickly applied to avoid a rapid escalation.[11]

The use of electrical devices, primarily the TASERS®,[c] has proven to be one of the most effective nonlethal force options and one of the most controversial. Like chemical agents, the most serious injuries from electrical devices[12] are nearly always second-order effects, but some observers have claimed that hundreds of people have died as a result of TASERS®. Nonetheless, these claims have been refuted by numerous studies.[13] For all practical purposes then, the focus for users needs to remain on avoiding second-order injuries. TASERS® work by inducing muscle **tetanization**, which is a medical term for involuntary muscle spasms. Understandably, a person who loses control of his muscles will have difficulty maintaining balance, which can result in injuries from falls. Although it's important to recognize that circumstances are seldom ideal for preventing injuries, some are so dangerous as to preclude the use of this type of force unless death is acceptable. Examples include, suspects in danger of falling from high places, into traffic, water, or machinery.

The only biological nonlethal option currently in widespread demand is the use of dogs in police and military operations. Most injuries resulting from the use of canines are first-order effects, especially puncture and crushing wounds to the extremities from teeth.[14] In fact, a German shepherd can exert a bite force of 1,500 pounds per square inch. Interestingly, only three documented deaths have occurred.[15] When it comes to injuries, canines have one major advantage over all other nonlethal options—they can increase the amount of force proportionate to the resistance of the adversary. Adversaries who attempt to fight or flee frequently suffer greater injury as the dog "regrips" to hang on or defend itself. Admittedly, it is an imprecise application of the principle

---

c Other devices exist; but the TASER® is, by far, the most popular and well known. Because the name is an acronym, it is capitalized. No further inference should be drawn.

but worthy of mention given the inability of any other option to adjust to resistance.

One of my partners, Capt. Richard "Odie" Odenthal, LASD (ret.), used to say that when it comes to nonlethal force, we "live with our enemy's wounded." This means that we are responsible for their safekeeping. After all, if we are willing to let them die, we can skip the intermediate options. Injuries of any kind create a huge logistical burden for medical treatment and evacuations and so provide tactical, as well as altruistic, reasons for avoiding excess. To be sure, any application of force, by its very nature, carries a risk of injury. An understanding of those factors that are likely to result in severe injury provides insight to decrease the probability and diminish the consequences.

6

Assessing and Managing Risk

THE USE OF force is always accompanied by risk. Assessing risk then, becomes an important factor in considering what force options will reliably accomplish operational objectives. Both the military and law enforcement communities are quite comfortable with identifying the criteria that justify lethal force.[1] After all, it's a threshold, and once crossed you can only be so dead. Nonlethal force, however, has no such threshold. Instead, there is a huge gray area where legitimacy and acceptability are blurry. For example, when considering how much injury is acceptable the question arises: Compared to what? Nearly all of us would accept substantially more injury if the alternative is death. Even so, we would likely consider nonlethal options that blind, disfigure, or maim as excessive even when the only alternative is deadly force. Furthermore, because the adversary remains alive, the situation may require that the force needs to be reapplied, or even escalated, before a satisfactory resolution. In the dazzling clarity of hindsight everyone can see the errors. Understanding the factors and influences involved in assessing and managing risk when employing nonlethal force options then becomes a critical capability.

Diversity Creates Difficulties

First of all, the human body is not consistently resilient. There are areas of the body where even a little force can cause grievous injury. When impact munitions are used, for example, striking any portion of the body where a bone is close to the surface—such as the head, cheeks, shins, elbows, knees, spine, etc.—compresses the soft tissue between the projectile and the bone, often resulting in lacerations. Likewise, human skin varies in thickness, from as little as .5mm for eyelids and the bridge of the nose to 4mm on the back and soles of the feet. As if that weren't enough, eyes and ears are so adaptive to excesses that any nonlethal option that attempts over-stimulation runs a serious risk of permanent injury. Not only is the human body not consistently resilient physiologically, we are not consistent psychologically. Our mental capacity to resist is highly influenced by things like surprise, fear, fatigue, pain, and anger. None of these factors is significant for the employment of lethal force, but each of them, and in any combination, directly affect the efficacy of nonlethal force.

Second, we differ as a population. Males and females, adults and juveniles, healthy and sick, strong and feeble, young and old, large and small, sober and inebriated, calm and agitated: all are factors in how safe and effective an application of some type of nonlethal force is.

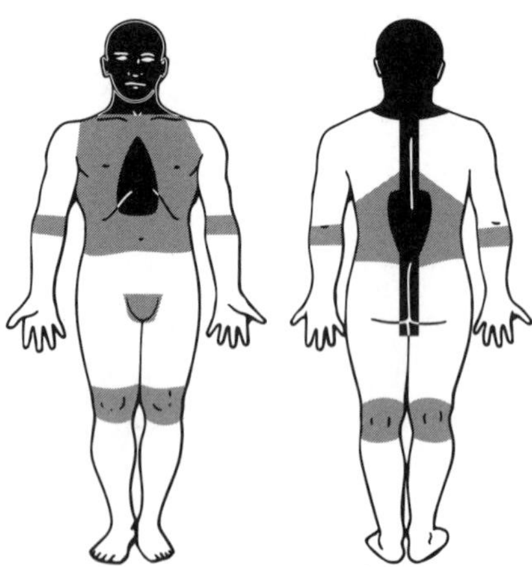

Figure 3: Vulnerability Areas of the Human Body

Impact munitions work by "pain compliance," meaning that they inflict pain to encourage compliance. While pain is fine, the associated trauma used to inflict it is problematic. The human body is not homogenous and there are areas of the body that are more vulnerable to injury than others and even mild impact can cause serious injuries. Coincidentally, the areas of the human body where most of the nerve endings are located also coincide with the more vulnerable areas to impact injury. This chart is typical of those used to teach police officers and military personnel the areas to avoid with nonlethal force options that require impact. The white areas are preferred, while the gray areas are to be avoided and the black areas are especially susceptible to serious, even lethal injuries.

This is one major difference between the military and law enforcement communities, since adversaries most likely to be encountered by a military force tend to be a lot more similar to each other than those encountered in law enforcement. With few exceptions, they are almost entirely male, of military age, in good physical condition, and healthy. Likewise, they tend to be mentally sound and psychologically resilient. Although some generalizations can be made for adversaries in law enforcement environments, they tend to be far more diverse. It is not at all unusual to find dangerous belligerents who are small, young, and female, or conversely, exceptionally large, older, and male. From personal experience, I can attest to the hazards when intervening with subjects displaying juvenile tempers and emotionally disturbed or insane individuals who are able to resist the harshest nonlethal force options, even when sustaining significant physical injury. No force option can safely and effectively accommodate all these differences.

Third, although people are certainly different, we are alike in many ways, too. For example, regardless of gender, size, age, and so forth, there are only four ways of **introducing** anything into the human body. These are by injection, inhalation, ingestion, or absorption. Whether it's tear gas, pepper spray, radio waves, or **electricity**, one of these avenues is required before it will have any effect on the human body. Likewise, our sense receptors are not evenly distributed. Our eyes, ears, nose, and mouth (light, sound, smell, and taste) are all located on one side of a relatively small appendage (more commonly called the face). These targets are quite small[a] and missing them completely negates some nonlethal force options. For example, even the most effective LED or laser aimed at the back of someone's head will be wholly ineffectual in dazzling a person's eyesight.

Finally, all but two commercially available nonlethal force options use pain compliance as their predominant means of effectiveness.[2] The general idea is to inflict enough pain that compliance is preferable to the suffering. There are, however, three major problems to this approach.

---

a On average, a human will have about twenty-two square feet of skin. However, the face is typically only about fifty-five square inches and each eye is only about one square inch! Because subjects trying to avoid the consequences are often moving and taking cover, precision marksmanship becomes exceedingly difficult.

First, pain is highly subjective. What is excruciating to one person can be only irritating to another.[b] Second, pain is not experienced equally throughout the body. The most sensitive areas, such as the face, hands, fingers, and groin, are also the most vulnerable to injury. To avoid excessive injury, we target the less vulnerable areas, such as the back, buttocks, and thighs. Third, and most importantly, there are few nonlethal devices[c] that can inflict pain without trauma. Pain has been used as a nonlethal option since antiquity. We are comfortable with pain as a mechanism to achieve compliance; it is the trauma used to inflict it that is troublesome.

Swett Curve

One of the best and most well-known methods of understanding risk with nonlethal force options is with the use of a **Swett Curve** (see Figure 4). The Swett Curve takes its name from Charles Swett,[3] who is credited with inventing it. Using a cumulative frequency distribution,[4] the Swett Curve compares force with the resiliency of the human population to identify three thresholds. The first is the one we've already identified—the amount of force that is lethal. The other two identify the amount of force that is effective and the amount of force likely to be permanently disabling or disfiguring. You'll note that each of the three threshold lines forms a lazy "S" shape. The lower tail of each figure identifies that portion of the human population that will be more affected by lesser force than the rest of the population. Typically, this has meant the very young, very old, and the infirm. Conversely, you'll note that the upper tails of each of the figures lean the other way. This is the portion of the population that is far more resistant to force than the rest of the population. This portion of the population tends to be large, healthy, determined males.

---

b Needless to say, the effects of pain can be greatly mitigated with subjects that are insane, emotionally disturbed, or under the influence of drugs.

c Although pain is experienced with both tear gas and a conducted electrical device, neither relies solely on it for effectiveness. Tear gas, especially pepper spray, causes blurred vision, a runny nose, and coughing, whereas a conducted electrical device uses muscle tetanization. Pain occurs; however, it is a byproduct rather than the intended consequence. Although not yet commercially available, the **Active Denial System (ADS)** uses radio waves to stimulate pain without tissue damage.

Ideally, nonlethal force options would be designed and employed so that they would affect the entire population beyond the threshold of effectiveness but short of being either permanently disabling or lethal. Given the technological limitations, however, nonlethal force options have hard thresholds. For example, regardless of whether the intended target is large or small, sick or healthy, weak or strong, a nonlethal projectile will strike with the same impact, a TASER® will discharge the same amount of current, and pepper spray will apply the same concentration. This means that there will always be a portion of the population that will be excessively affected and another portion that will be minimally affected. Recognizing this problem, manufacturers of nonlethal weapons and munitions almost

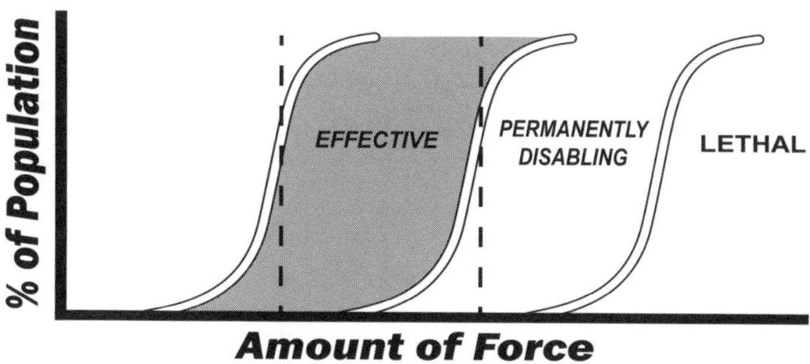

Figure 4: Simple Swett Curve

One of the best methods of estimating the amount of force for a given population is the so-called "Swett Curve." Each of the "lazy S-shaped" curves estimates the amount of force necessary to be effective. The lower ends include the very young, small, infirm, or injured members of the population, who will respond with a small amount of force. On the upper ends are the fit, strong, determined, emotionally disturbed, or intoxicated members of the population, who require a much greater amount of force to achieve the same results. The first curve identifies the minimum level of force for effectiveness; the second is the threshold at which that same force becomes dangerous and even permanently disabling; and the third identifies the threshold where it ultimately becomes deadly.

Ideally, a force option would be effective without causing serious injuries (as indicated by the gray area). The problem is that no force option is currently available that can adapt to the vulnerabilities and resiliencies of the human population (as indicated by the vertical dashed lines). As can be seen in the diagram, the same force option may result in vulnerable members of the population suffering permanently disabling injuries, while being ineffectual against the more resilient members.

Figure 5: Dangerous Nonlethal Force Option

A dangerous force option is one in which the permanently disabling threshold is close to the effective threshold. Examples include those options that affect sight and hearing, which are so resilient to over-stimulation that it is difficult to achieve effectiveness without becoming dangerous. This illustration reveals that it still may be nearly impossible to kill someone, but that the option being considered is still dangerous because of the close proximity of the threshold for becoming permanently disabling.

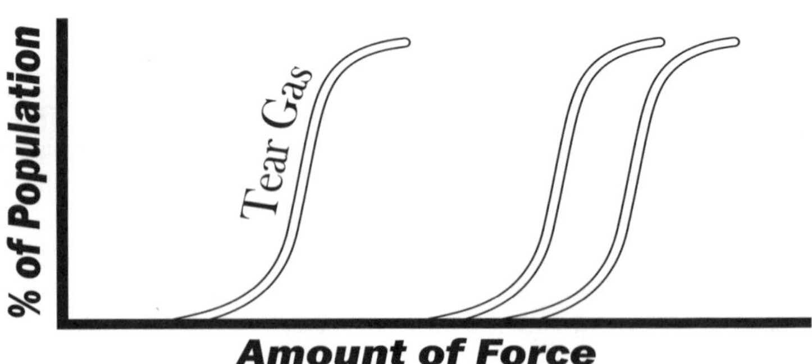

Figure 6: "Forgiving" Nonlethal Option

Some nonlethal force options have been determined by both science and experience to be extremely safe, while remaining effective. The use of CS (tear gas) as a riot control agent is an excellent example. While effective with a low dose, it remains relatively safe at more than a thousand times that dose. When it becomes dangerous, however, it is also close to becoming lethal, which is illustrated in the graph by the permanently disabling threshold being close to the lethal threshold at the right-hand side of the graph.

always err on the side of safety rather than effectiveness.[5] This means that a greater portion of the population will be under-affected than over-affected, to avoid causing severe injury.

While the figures here are only for illustrative purposes, it is simple to estimate how dangerous a nonlethal option might be because the proximity of the three curves to one another is directly related to the type of nonlethal force being examined and how dangerous it is. For a device that uses light or sound stimulation, the effective threshold lies very close to the permanently disabling threshold, indicating the small safety margin when attempting over-stimulation without harm, while the lethal threshold remains quite far apart[6] (see Figure 5). There is little margin for error and these devices can be dangerous even when properly used.

If we examined pepper spray or tear gas, the thresholds for permanently disabling and lethal would be closer together but quite far away from the effective threshold indicating the much wider margin of safety. Nonlethal options with wide margins of safety are called "forgiving" (see Figure 6).

As can be seen, the Swett Curve is one of the most useful tools for comparing risk, even with the disparate nonlethal options that may take a wide variety of forms such as liquids, sprays, projectiles, chemicals, electricity, directed energy, and so forth.

While the Swett Curve makes these issues clear for planning and policies, it provides little guidance for those who are actually engaging combatants. A simpler system is needed. For impact devices,[d] law enforcement often uses a three-color system based upon traffic signals. Portions of the body where injuries tend to be relatively minor and temporary are targeted and colored green, while areas likely to result in serious and lasting injury or even death are colored red and areas where injury tends to be moderate and enduring are colored yellow.[e]

Of note is that very few law enforcement agencies use an evaluation system that prohibits the use of a particular nonlethal option altogether.

---

d More commonly referred to as **kinetic** devices in the military community.

e From long experience in the field, it appears that where a person is hit has a greater influence on the degree of injury than what type of nonlethal option was used (see Figure 3 on p. 42). This is one topic ripe for further study.

Police officers are always expected to use good judgment and only that force which is reasonable and no more. Prohibiting an option altogether leaves the problem unresolved but the means to conclude it are limited to other options that are always harsher and more primitive than that which has been prohibited.[f] As they say, "Better decisions are made on scene than on paper."

Understanding Risk

A thorough examination of **risk** reveals that it has two fundamental and integral attributes, **risk probability** and **risk exposure**. The relationship is sometimes expressed as risk equals probability plus exposure: R=P+E. Changing either the probability of the event or the exposure to harm will also change the risk.

The first attribute is probability, which is an estimation of the likelihood of some misfortune or setback. It is expressed as either a percentage or a ratio. One example to illustrate this concept with nonlethal munitions is a metric used for impact munitions by manufacturers and developers, which gives the likelihood of hitting a "man-sized" target at a given distance.[7] A developer might say that their munitions could hit a man-sized target 80 percent of the time at forty yards. This measurement of accuracy means that given the range prerequisite, it will reliably strike an adversary four out of every five shots. The probability then, is 4:5 or 80 percent. Conversely, it also means that shooting at rioters in a mob could result in the injury of a person (other than the intended target) one in every five shots. Thus, the failure rate is one in five or 1:5.

The second attribute is exposure. Exposure describes the risk of adverse consequences resulting from being susceptible to harm. Using the same example, it would describe the likelihood of missing the intended target and/or unintentionally striking a bystander. Typically, people will take precautions or even avoid situations where they may be injured. Thus,

f It should go without saying that if an alternative is available that is as safe and effective as the one prohibited, it would be preferred in all cases. Without exception, however, nothing in tactics comes without a tradeoff. Something is given for anything gained.

reducing risk through exposure may only require informing potential victims of the consequences.

When assessing risk, both probability and exposure are important and decreasing either will reduce the risk. Although it is often difficult to change the probability, it is far easier to change the exposure. This is often done by limiting the use of a nonlethal force option in the rules of engagement: for example, by requiring closer ranges or by separating noncombatants from combatants. The chance for a failure, then, is compensated for by changing the exposure. For example, if we know that the chances of missing our target is one in five (which is also the risk of hitting an adjacent bystander), we can restrict the use of a particular option when others are in close proximity or by isolating combatants. Unlike probability, which can never be reduced to zero, eliminating exposure makes it possible to avoid the adverse consequences for specified individuals altogether.[8]

It needs to be mentioned that the likelihood of failure (probability) with nonlethal weapons can be attributed to many things—like marksmanship, maintenance, calibration, and the like—but is also a problem of design, manufacture, quality control, and so forth. Accordingly, the predominant responsibility for reducing risk by decreasing probability lies with developers and manufacturers. Exposure, however, is always situational and the responsibility of decreasing risk by reducing exposure lies entirely with the users of nonlethal force options.

I would very much like to conclude with definitive guidance and algorithms that can be used to accurately and reliably predict the various risks associated with nonlethal options. Unfortunately, there are simply too many uncontrollable variants. Some years ago, I worked with a PhD from the USAF on a study for determining a "battle damage assessment"[g] methodology when employing nonlethal weapons in military situations.

g For you law enforcement types, a battle damage assessment (BDA) is an estimate of damage resulting from the application of force, either lethal or nonlethal, against an objective. For example, if you drop a 250-pound bomb, it will throw shrapnel out to "x distance" and is likely to cause wounds out to "y distance" and is likely to kill out to "z distance." It is used in planning for selecting a weapon, to how many repetitions, to how many will be needed for a given effect.

He wanted to know the effects of various nonlethal projectiles for estimating everything from inventory and cost to resupply and predictable effects. He became somewhat impatient when I explained how difficult (i.e., impossible) that was going to be and pressed me with the questions:

> Dr. S: "Let's take a typical stun bag round fired from a shotgun. What will happen to the adversary?"
>
> Sid: "What was the range? Where on the body did it hit him? What was he wearing? Was he under the influence of anything? If it knocks him down and he gets back up, does it still count?"
>
> Dr. S: "Do all those things matter?"
>
> Sid: "Alone, and in any combination."

Needless to say, this problem remains unsolved. This is why a thorough understanding of the factors and influences involved in employing nonlethal options is so critical. It is why knowledgeable planners and decision makers are so crucial. It is why, no matter how pure and noble your intentions, how meticulous your plans or how diligently and carefully you employ them, bad things will still happen.

7

Voodoo Science and the Media

ONE MIGHT EXPECT critics of injuries and deaths resulting from military and police confrontations to be among the first to laud the efforts to develop and employ safe and effective nonlethal options. Such has not been the case, however. Nearly from their inception, nonlethal weapons have been fraught with controversy. The extensive use of baton rounds used by British forces in Northern Ireland generated intense pressure to abandon them altogether. Later, the use of CS gas was criticized. Sensational headlines throughout the United Kingdom both lauded and condemned the use of nonlethal weapons, especially those being used in Northern Ireland. In fact, the most controversial aspect of policing in Northern Ireland has been the weaponry used by police.[1]

Similarly, controversies involving the use of various nonlethal options occurred in the United States, especially the use of pepper spray and TASERs®.[a] Since the early 1990s, claims have been made in the United States that people have died unnecessarily after being exposed to tear gas and pepper spray. In 1999, the American Civil Liberties Union (ACLU) claimed that scores of people were dying as a result of pepper spray and that it was being used as a form of punishment and torture comparable to a "chemical cattle prod."[2] A series of lawsuits attempting to ban or limit the use of pepper spray followed. News reports repeated sensational claims that it appeared that "one person dies for every 600 times pepper spray is used."[3] Studies that repudiated such claims from the National

a Although I'm sure there are those who prefer a generic term like "conducted energy device," in this case they were all TASERs®.

Institute of Justice and the International Association of Chiefs of Police (IACP) were slow in coming but provided strong evidence of the dubious, or at least highly exaggerated, nature of these claims. Follow-on studies also supported the use of pepper spray as an effective and relatively safe method of controlling violent persons, especially when other alternatives were inappropriate or unavailable. As the evidence continued to accumulate, the controversies, and related news value, diminished.

While much of the controversy on the use of pepper spray has died down in the United States it has been supplanted with the popularity of the TASER® electronic control device. While TASERs® have been around since the 1970s, newer versions became available to law enforcement in the late 1990s.

The employment of the newer TASERs® produced strongly encouraging results, often rating effectiveness well into the ninetieth percentile;[4] unheard of for any force option—even lethal force. Within months of the first TASER® uses, controversy erupted when Amnesty International reported the death of people after being TASER®-ed. Since then, the controversy has only heightened, with Amnesty International now reporting the death of hundreds of people as a result of "TASER®-related" deaths.

So, what is the true story? Why are caring and well-meaning organizations, like the American Civil Liberties Union and Amnesty International, convinced that TASERs® and other nonlethal options are so dangerous? Why do news media outlets, who pride themselves on their objectivity, report findings that are inaccurate or biased and are later refuted? More importantly, how does a government make informed decisions about the weapons and force used on their behalf when they receive conflicting information? Herein lays the crux of this chapter.

Simply put, much of the "science" cited by detractors of nonlethal options is either fundamentally flawed, misunderstood, mischaracterizes the evidence, or ignores influences beyond the control of the user. It is an exceedingly rare occurrence when any of this so-called science is subject to peer review by objective researchers, cited authoritatively in reports by other scholars or scientists, used in supporting references, or corroborated with other research by bona fide scientific or academic institutions.

Although it may take on any number of forms, it can be described as "voodoo science."[5] Voodoo science is a catch-all term for any junk science that misleads or mischaracterizes the evidence. It matters not whether it was intentional or not, only that on the surface it is predisposed to make people believe something without scientific validity. While voodoo science can take on any number of forms, four types are used so often in describing nonlethal weapons that they merit mention.

Pseudo-Symmetry

One of the most common types of voodoo science in the media results from "**pseudo-symmetry**." This occurs when even well-meaning news media attempt to obtain a balanced outlook by seeking out and quoting opposing views. Although this seems both logical and reasonable, it becomes misleading when an opposing view is actually held by a very small minority. Because the two opposing views are presented with equal emphasis it appears to viewers and readers that subject matter experts are equally divided on the issue when, in actuality, the vast majority may hold a differing opinion. As can be imagined, pseudo-symmetry presents unparalleled opportunities for small minorities with radical viewpoints.

In the controversy surrounding nonlethal options, one example of pseudo-symmetry occurred in 2002, after the Dubrovka Theater Siege in Moscow, Russia. This was the first use of **soporific** chemicals as a nonlethal option. Even though the employment of this nonlethal option saved the lives of hundreds of hostages, 129 others died as a result. In the media frenzy that followed, a previously unknown group called the Sunshine Project rocketed to national prominence when they were called upon to provide comment by the news media. Prior to this event the group had only existed for about thirty-six months, had a staff of fewer than five people, and at its height had fewer than 1,300 followers on their largest Internet distribution list. It is now defunct.[6]

Ignoring or Distorting Perspective

Ignoring or distorting perspective is another characteristic of voodoo science. This occurs when an emotion-arousing factor is sensationalized without a contextual comparison. One of the best examples of ignoring perspective is the current controversy in the United States concerning the use of TASERs®. Civil libertarian groups and some media organizations contend that more than a thousand victims have died as a result of "TASER®-related deaths." Although rarely do independent post-mortem investigations specify TASER® exposure as a cause of death, or even a contributing cause of death, for the sake of argument let's assume *all* are causally related to TASER® exposure. A valid scientific study would ask questions like, "How many people were exposed to TASERs® that didn't die?" or "How many people died in similar circumstances but were not exposed to TASERs®?"

This is a case of being "attacked by the numerator." The answer to these questions provides a denominator. To gain a true perspective of the dangers involved, both the numerator and the denominator must be known. The latest available figures report 1,081 deaths in the US after being shocked with a TASER®.[7] That is the numerator. The estimated number of suspects exposed to a TASER® is 3,933,054.[8] All that is then required is to divide the numerator by the denominator to reveal that the chance of death is about .03 percent!

But wait! There is another group besides suspects that are exposed to TASERs® because police and military students routinely volunteer to be hit with a TASER® to personally experience the effects. These statistics are never reported. As of this writing, there have been at least 2,559,343 members of this group, and more significantly, this group has never experienced a single death![9] Thus, the combined size of both groups (the entire population of people who have experienced a TASER® shock) is 6,492,397 and the chances of death are about 0.016 percent![10] Accordingly, even if *every* post-TASER® death was a direct result of exposure, the likelihood of dying following being shocked with a TASER® would be less than two-hundredths of one percent. Certainly, any death is a tragedy, but by ignoring the total number of uses it appears as if the event is far more common than it is.[11]

Anecdotal Accounts vs. Statistical Evidence

Because of our faith in science, any evidence that appears to be scientific is automatically awarded more credibility than other types. Accordingly, a third manifestation of voodoo science in swaying opinion is by gathering anecdotal accounts, especially those that seem egregious, and presenting them as evidence. For example, listing newspaper articles, citing civil court cases, or quoting litigants and plaintiff's attorneys have proven very successful in capturing the attention of an audience. Nevertheless, given that newspaper accounts are nearly always second-hand, anecdotes from a reporter quoting a witness and relying on statements provided by people who have an obvious interest in an outcome would be viewed individually as suspect. However, when taken together they appear to corroborate each other. This type of voodoo science may best be described as pseudo-science, since its practitioners may believe it to be science but to objective researchers it is flawed on its face. Of the various forms of voodoo science, this type is second only to pseudo-symmetry in prevalence, especially when supporting opinions for reports and press releases.

Fallacies

The fourth of the most common types of voodoo science is when fallacies are used. A fallacy occurs when an argument or statement is supported by an invalid inference. In the case of nonlethal weapons, it is most commonly manifested by either ignoring the influence of other factors or making assumptions that exceed the evidence. No better example exists than in the claims of hundreds of "TASER®-related deaths." Without ignoring the possibility that TASERs® might have a role in the deaths, it is disingenuous not to consider that many of the people who have died after being shocked with a TASER® had serious, preexisting life-threatening medical issues, many times because they "self-medicated" with cocaine, PCP, methamphetamine, or other drugs. Furthermore, many were engaged in life-threatening behaviors that required some type of intervention to protect human life—including their own—and without an ability to employ an effective nonlethal option they might very well have been killed outright. In these circumstances, even the noble attempts to save their lives are made to appear contemptible.

Collaboration vs. Competition

Military and law enforcement organizations, as well as civil libertarians, seek safer and more effective methods of reducing death and injuries during violent confrontations, but are so at odds in how best to achieve these reductions that more effort is invested in arguing than in cooperating. It is particularly troublesome, however, when even the supporting information is so tainted as to render it meaningless.[12] Certainly, voodoo science is not limited to the controversy surrounding nonlethal options. In recent years, the public has been terrified of cancer from cell phones and captivated with the possibilities of unlimited, pollution-free energy from cold fusion, neither of which has withstood the scrutiny of scientific inquiry.

Nonlethal weapons can provide moral alternatives in inherently amoral circumstances, but completely safe and effective technologies remain elusive. In the search for these options, only objective approaches to examining them can balance the risks and benefits of their use.[13] Protecting the peace while preserving life is a noble calling, but it is far more difficult to achieve without objective science and fair reporting.

8

The Search for the Magic Bullet

THE SEARCH FOR more effective nonlethal options has gained more momentum in the last three decades than in all previous history. Law enforcement agencies across the globe encounter each day situations that justify the use of lethal force, and so *any* nonlethal alternative is appealing. In domestic law enforcement applications, this situation has resulted in nonlethal alternatives that are immature or have side-effects but that are still fielded. There is no better example of the desperate desire for better options, because even with little or no testing they are still preferable to the pitiless nature inherent in lethal force options. This has created an ideal market for the development of nonlethal options, since, as was mentioned in the Introduction, the standard is not perfection, it is the alternative. And the alternative is usually deadly.

On an even grander scale, innumerable contentious social and political issues ranging from pollution and global warming, to the reunification of Korea or Palestinian nationalism, have erupted with violent protests and so created a global impetus to find options short of deadly force. The need has been so desperate in some circles that it has been compared to the search for the Holy Grail. Notwithstanding any altruistic motivations, any developer who finds a nonlethal technology that is even moderately effective while remaining truly nonlethal will immediately become wealthy beyond their wildest dreams.

In the field of nonlethal technologies, the common expression is the search for the **magic bullet**.[1] It is not known when or where the term "magic bullet" was first used to describe a truly effective and completely nonlethal solution. It is based upon a metaphor for any straightforward solution, especially some new technology perceived to have exceptional effectiveness. Its roots have been attributed to both the silver bullet in folklore effective against werewolves and such, and the silver bullet used by the main character from the radio and television *Lone Ranger* series.

Undermining this impetus, however, are two powerful countervailing influences. Both deal with the risk accepted in employing a nonlethal option, which admittedly, is always less than perfect and often less than effective. The first is the risk in being first. Given that every nonlethal device has some drawbacks, not the least of which is that none can claim with absolute certainty that they are always nonlethal, they will be subject to criticism and civil attack for a failure. This is especially troublesome with new devices, which are not only intensely scrutinized but are more prone to failures and shortcomings than mature technologies. This latter issue is particularly problematic for the law enforcement community in that the pathways through the legal quagmire are well charted for lethal force but fraught with pitfalls for nonlethal options. Nobody wants to be the one to "bell the cat."[2]

A second problem is termed the "revenge factor." This condition was first identified by American sociologist, Dr. David Klinger.[3] After studying a number of tactical operations in which nonlethal devices were used, Dr. Klinger noted that because of their primitive nature in comparison with lethal force, the person who attempts to use them also accepts the risk that they might not work and result in the death of either the adversary or the user. More poignantly, persons who elect to employ nonlethal options can find themselves in a position of attempting to spare the life of the person trying to kill them.[4] A paradox is revealed because the application of a nonlethal device can result in what it is trying to prevent.[5] Together, these counter-influences present formidable challenges and are often enough to dissuade even the most intrepid commanders from accepting the additional risk required to employ nonlethal options.

Criteria

In deciding what the magic bullet might look like it is useful to identify some meaningful metrics. The most critical one is that truly effective nonlethal options must provide adequate protection against their lethal counterparts. Any force option that fails to meet this high standard will not provide adequate advantage for universal application[a] and so will always be reconciled as a "nice to have" rather than a "need to have."

A second requirement is that the option must be affordable. Nonlethal force options are subject to the same market constraints as any other technology and even an ideal device that is prohibitively expensive will not gain widespread approval. Generally, this means that a nonlethal device must be priced competitively to the lethal device it is attempting to replace.[4]

I'm not so presumptuous as to assume that I have any more insight into what the magic bullet will eventually look like than the learned scientists who are diligently seeking some solution, but having personally benefitted from the Marine Corps' tour of developing countries in crisis, not to mention my years in law enforcement on the streets of Los Angeles, I have developed some rather strong opinions. It would then seem appropriate to provide what guidance and direction I can:

- First and foremost, the device must have universal application. For whatever reason, any nonlethal device that cannot be used inside buildings; in rain, darkness, stiff winds, or cold weather; or against smaller people, is not, by definition, completely effective.

- Second, it must be discriminating. This means that the effects must be precise enough to affect the intended target and *only* the intended target. Any device that cannot exclude innocent bystanders may preclude it from being used whenever they are present. Such a device would not be practical.

- Third, it will need to be environmentally benign. This means that it must not only not alter the environment, but it must remain safe and effective in altered environments. The use of conventional riot control agents, for example, alters the environment and requires harsh "tradeoffs." For that reason, the use of tear gas

---

a And there are none to date!

is often forbidden near schools, hospitals, shopping centers, airports, major intersections, and the like. Conversely, devices that are sources of combustion, such as flashbangs, stingballs, and many launchable projectiles, cannot be used in altered environments that may contain flammable vapors or liquids, such as drug laboratories, passenger aircraft, factories, and so forth.

- Fourth, it must be highly portable. Even the most effective device will have little value if it cannot be easily and quickly employed. Situations that will most likely benefit from the use of nonlethal options are highly volatile and can quickly escalate to a point where nonlethal options are no longer adequate. If it is not available when you need it, it doesn't matter how well it works. Frequently in military situations and nearly always in law enforcement situations, this means man-portable.

- Fifth, it must be reusable. This means that the device must remain serviceable and usable on more than one suspect or multiple times on a single suspect.[b] This may require multiple projectiles, charges, applications, sprays, or whatever method employed, but the critical aspect is that it must not commit the user to a single course of action by a failure.[c]

- Sixth, the effects must be completely reversible. This means that a person affected by a nonlethal force option will completely recover with no aftereffects. Thus, the device can neither cause

---

b Admittedly, this requirement may upset certain critics, since they suspect that people like me will use it for torture. In point of fact, there are just too many things in a real-life situation that we don't control. We've had TASER® darts ricochet harmlessly off the snaps on a windbreaker, sudden gusts of wind divert streams of pepper spray, and stun bags that struck a paperback in a pocket, rendering them completely ineffective. To be sure, these are exceptions, but they are not so rare or exceptional that those of us inheriting the aftereffects will ever discount them.

c This is one of the reasons why directed energy weapons are so popular. Called the "deep magazine" concept, it refers to the ability of repeated, even indefinite, applications as long as there is power available.

permanent injury, nor have lasting medical problems such as disease, mutations, scarring, permanent discoloration, etc.

- Seventh, the effects must be nearly instantaneous. Generally, this refers to human reaction time—about three-quarters of a second. The use of any nonlethal force option that allows time for a suspect to have a reaction before becoming effective must then consider a counteraction by the suspect. This increases the complexity of the problem logarithmically because the suspect's counteraction must then also be considered as part of the reasonable and likely consequences.[d] And, because our counter-counteraction then becomes a factor, the problem becomes infinitely difficult.

- Eighth, it must be a hundred percent effective. This means that it must prevent or incapacitate rather than just debilitate. Currently, the only nonlethal option available that even approaches a level that might be considered incapacitation is the TASER®, and even these have had many instances in which a belligerent has remained functional even with repeated exposures. To be a hundred percent effective a nonlethal option must render a belligerent either physiologically or psychologically incapable of resisting.

- Ninth, it must be completely safe. This means that any application, or repeated applications, must work equally well on the 325-pound professional body builder and the 100-pound high school student without increasing the likelihood of injury. Likewise, it must be safe for all genders, sizes, ages, races, weights, and medical conditions.

- Finally, it must be effective at ranges that do not require undue risk on the part of the user.

---

d The application of force is always unpleasant, and belligerents will go to great lengths to avoid it, as well as to negate the effects whenever possible. One means can be as simple as to slide a thick magazine inside the shirt to nullify the pain from impact munitions or to wear safety glasses to fend off pepper spray.

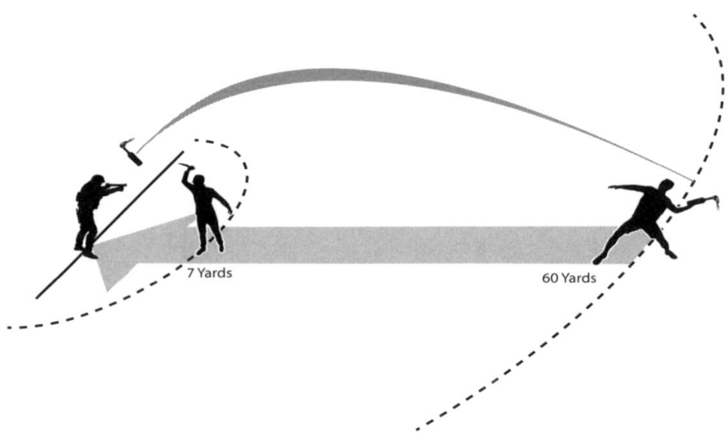

Figure 7: Critical Vulnerability Ranges

Understandably, any effective nonlethal force option must be able to provide adequate protection against assailants. Typically, range is the predominant factor and two are especially important. Seven yards is the range commonly accepted where an assailant armed with an edged weapon or a club could inflict fatal injuries, if the force used to deter him is not immediately effective. Less than three percent of the human population can throw an object large enough to cause serious injury or death beyond sixty yards.

Critical Ranges

Depending on the circumstances, there are generally two ranges that provide reasonable thresholds of safety. The first range is twenty-one feet (seven yards). This is the range considered to be potentially lethal if the force used to deter a combatant armed with an edged weapon or club is not immediately effective.[7] Consequently, any nonlethal option that falls short of that requirement means that the user is forced to accept the consequences of failure when employing the nonlethal force.[e] The second range is sixty yards (fifty-five meters). (See Figure 7.) This is because only about three percent of the population can throw an object large enough to cause severe injury[8] beyond this range.[9] Like the twenty-one-foot rule, any

e For any of you hearing this for the first time, it is also why a lot of the suspects armed with knives and clubs don't survive the encounter.

nonlethal option employed with the intention of preventing rioters from hitting the user, or nearby personnel, will require a range of sixty yards or more. It is equally important for adversaries because without a nonlethal option capable of being effective at ranges of sixty yards or more, rioters attacking with Molotov cocktails (a.k.a. gas bombs or petrol bombs) are just as likely to be engaged with lethal force.

The De Facto Standard

Somewhat surprisingly, these requirements are well known to the public and have become a de facto standard against which all available nonlethal force options are compared. This remarkable phenomenon was first observed on a Thursday night in September of 1966. For three years afterwards, the public was treated to the use of the ultimate nonlethal force option in the form of the phaser on the television series *Star Trek*. There is hardly a person in the western hemisphere who can't describe the appearance, sound, and effects of a "phaser on stun." Admittedly, there has never been an actual phaser, even in a primitive form, and the fact that the script is set three hundred years in the future makes it truly a development solely residing in science fiction. (See Figure 8.)

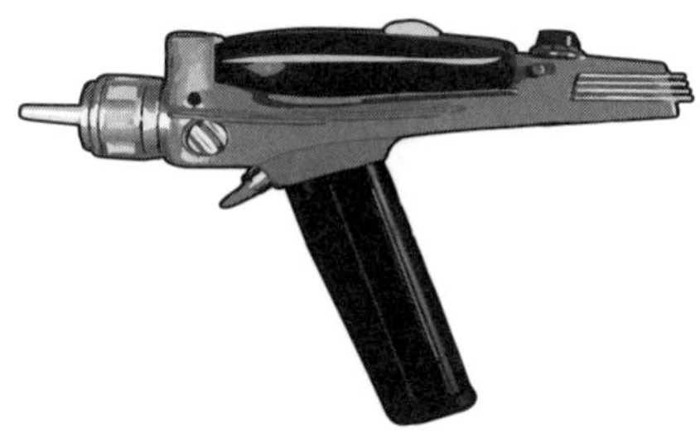

Figure 8: The De Facto Standard

Ironically, even though it does not yet exist, nor is it likely to be developed in the foreseeable future, the standards for which the ultimate nonlethal weapon will be judged have been identified. Since the conception of the "phaser," which first appeared on the television show *Star Trek* a half-century ago, the public has come to compare all nonlethal weapons with it. Whether we like it or not, there is scarcely a person who has ever seen its TV use that can't describe the sound, effects, and expectations for the nonlethal effects and it has become the de facto standard for a truly safe and effective nonlethal force option.

That said, the public did readily accept the phaser as the ultimate nonlethal force option, and in an era of scientific advancement it seems not so out of reach. To be sure, these standards based on a device that has never existed are unrealistic, but that doesn't mean that they can be ignored. As former US Attorney General Janet Reno said in testifying before Congress after the Branch Davidian fiasco: "If we can send a man to the moon, we ought to be able to develop nonlethal technologies. . . ."[10] As frustrating as the search has been, notable progress has been achieved in recent times, because for the first time in history, the need exceeds the risk.

9

Classifying Nonlethal Options

THERE IS CURRENTLY no universally accepted taxonomic system for nonlethal weapons, consequently descriptions and terminology will differ from country to country, agency to agency, and even region to region. As just two examples, so-called rubber bullets are all projectiles but have been classified by different institutions into categories variously described as, impact munitions, kinetic munitions, specialty impact munitions, blunt impact munitions, kinetic energy impact projectiles, and extended range impact munitions. Likewise, the electrical stimulus (TASER®-like devices) have been placed in similarly confusing categories such as, conducted energy devices (CED), electronic control devices (ECD), electro-muscular incapacitation devices (EMID), electro-muscular disruption devices (EMDD), neuro-muscular incapacitation devices (NMID), human electro-muscular incapacitation devices (HEMI), neuromuscular incapacitating device (NID) and electrical weapons (EW). As with the other nonlethal terms, there does not appear to be any consensus likely in the foreseeable future.

Central to a shared understanding of any complex subject is a common language. No real understanding can take place without it. Doctors, pilots, carpenters, financiers, police, and military all use terms unique to their particular professions that have precise meaning among their members. It goes without saying that this problem with nonlethal force tremendously complicates precise descriptions, meaningful comparisons, and clear understandings. Collaborative efforts are nearly stymied until a common language is established and agreed upon. Nearly without exception, every report and study on nonlethal weapons is accompanied by a glossary identifying the terms used and their meanings. The problem has long been recognized, however, and attempts have been made to develop a useful taxonomic system for force, most notably by the International Law Enforcement Forum (ILEF) and the US Department of Defense's Joint Non-Lethal Weapons Directorate.[1] Neither has gained much acceptance beyond their own circles of influence and confusion remains widespread.

Having personally participated in many projects and studies to bring some closure to this issue, I don't pretend to be able to provide anything better. That said, it would seem prudent in a book of this nature to provide readers with at least some understanding of the current state of affairs, and the supporting thoughts and obstructions. Accordingly, the following is loosely based upon discussions with many experts on this problem.

Nonlethal Taxonomic System

In order to be useful, a taxonomic system needs to not only describe the various items but provide the rationale for assigning new items. Accordingly, one of the most important steps is to identify these conventions and the following have been suggested by various nonlethal working groups. The remainder of this chapter is my attempt to amalgamate the best thoughts from the various nonlethal working groups dealing with this issue rather than any one group. I have no hopes that it will be adopted, only that we have to start somewhere, and these represent the thoughts of the best and brightest minds familiar with this field:

- The first requirement is that, to the greatest extent practical, the system should use those terms that have gained even minimal acceptance rather than prescribe new ones. This concept is called "**standards by consensus**" as opposed to "**standards by mandate**." This requirement is necessary because not only is there no regulatory authority to require it, once a term has become popular it is nearly impossible to change.

- The second is that considering the different cultures, jurisdictions, disciplines, and languages, the system needs to be simple enough to be understood by everyone. Once again, because there is no recognized authority, everyone involved will need to readily understand how such a system works, for it to be accepted.

- The third is that assignments should use those individual characteristics that distinguish one option from another and are conspicuous enough to allow easy identification and avoid confusion. These characteristics are described later in this chapter.

- The fourth requires the system to be infinitely expandable to accept new innovations without requiring a change in the methodology. Consequently, should a new type of weapon or munition become available that will not meet the conditions of the existing hierarchy, an additional grouping can be added without changing the definitions of the current ones.

- The fifth is that, to the maximum practical extent, the system needs to be exhaustive. There should never be a necessity for catchall terms like "other" or "miscellaneous." This would only perpetuate the very problems such a system is designed to solve.[2]

The rudiments of such a system have been roughly outlined by many organizations and special interest groups but, to my knowledge, have never gained acceptance beyond the agreements of those who participated. With that in mind, let these criteria serve as a foundation for the thoughts that such a system would need to expand on before becoming practical. Following these protocols, breaking the force domain into manageable chunks would then proceed as follows.

Type

The first separation is by type of force. Types are separated by intent. Whether a force option is intended to be lethal or not is the definitive criterion. In the force domain there are only two types, lethal and nonlethal. Because intent is the distinguishing criterion, how a weapon is *used* is the defining characteristic rather than how it is *designed*. Consequently, a lethal weapon used to fire a warning shot would be assigned to the nonlethal type of force. Conversely, a nonlethal weapon used with the intent and manner of causing a life-threatening injury or death would be assigned to the lethal type.[3] The remainder of this chapter will focus exclusively on nonlethal force.

Role

The second separation is based upon their predominant role. This grouping describes how a particular nonlethal force option is used. Generally speaking, there are five roles for nonlethal force:

- The most well known role for nonlethal force options are those predominantly designed as anti-personnel. That is, restraining individuals from doing something. Devices designed to achieve this goal act either directly on an individual or indirectly through the use of barriers or area denial of some type. Because anti-personnel devices can be used to prevent a person from driving (**anti-mobility**) or entering or escaping from an area (area denial) they are the most versatile of all the classes.

- The second role is in anti-mobility. These devices are designed to prevent the use of vehicles, aircraft, vessels, and other modes of transportation. Second only to safe and effective anti-personnel options, this capability is the most sought-after nonlethal option in the world, especially against privately owned vehicles, which law enforcement agencies across the globe chase on a daily basis.

- The third is area delay or denial, or less commonly referred to as "anti-access" options. These force options are intended to inhibit or prevent passage through, or access to, an area. While this would seem to be more important to the military community, it is equally important to law enforcement, who are responsible for incarcerating felons in prisons and the accused and sentenced in jails. Similarly, an ability to deny belligerents access to areas like schools, playgrounds, airports, military installations, freeways, and the like, without relying on lethal force, has obvious advantages in both military and law enforcement applications.

- The fourth is anti-materiel. These options attempt to render useless those necessary supplies and support essential for warfighting. Obviously, this role is of more concern to the military than law enforcement and includes options that render useless such things as weapons, equipment, vehicles, fuels, and provisions.

- The fifth role attempts to affect an entire infrastructure. These involve degrading or preventing the use of basic features or facilities serving a large segment of a population. Power, water, communications, manufacturing, finance, commerce, and mass transportation are common examples of functions that could be targets to achieve this objective. Understandably, this is of nearly no interest to law enforcement but has many advantages in military operations.[4]

Category

The third grouping is by category. Categories are distinguished by the predominant method of operation to achieve the desired effects. To date, there are six categories of nonlethal options:

- The first is those that work by impact. These are also referred to as "kinetic" options, especially by the military. Nonlethal options assigned to this category require an impact to function. Whereas nearly all in this category induce pain, some are designed to burst upon impact to disperse a liquid or powder. This is just one example of the difficulties in classifying these options, since a munition fired at a wall to release a chemical agent needs impact to function, but the effects are chemical rather than kinetic.

- The second category is chemical agents. Chemical agents include any that use a substance to create a pharmaceutical effect. Riot control agents and tear gases[a] are probably the best-known nonlethal chemical agents, but as **calmatives**[b] and **malodorants**[c] become more available, they would also be assigned to this category.

- The third category is mechanical devices. Mechanical devices are those that restrain the intended motion of something. Whether used on people, vehicles, aircraft, or vessels, these entanglement systems take many forms, such as nets, **adhesives**,[5] **lassos**, **loops**, or hooks. Conversely, **super-lubricants** work to make it difficult to grasp objects or even stand on hard surfaces. Described as "slicker than snail snot," these nonlethal slippery slimes have been in development for at least three decades and can be dispersed as dry granules and activated with water.

- The fourth category is electrical. True to the name, this category is reserved for nonlethal options that use electricity as their predominant method of operation. The best known are almost certainly the TASERS®, but would also include **shock belts**, **electric batons**, and the like.

- The fifth category is biological systems.[6] This category comprises nonlethal options that are alive. The best example is certainly the use of canines,[7] but would also include horses, stinging and biting insects, noxious plants, microbes, and so forth.[8]

- The sixth category is directed energy. Directed energy includes devices that work by emitting any type of energy over free space. Current devices include those that emit light, sound or radio waves. Contemporary examples include the Active Denial System (ADS), which uses **electromagnetic energy**, a laser dazzler (light

---

a The most common examples are 2-chlorobenzalmalononitrile (CS) and oleoresin capsaicin (OC), generically referred to as pepper spray.

b Calmatives refer to chemicals that work as a **sedative**, **hypnotic**, analgesic, **somnolent**, soporific, and so forth.

c Malodorants are chemicals that work by giving off a potent stench.

energy), and the **Long Range Acoustic Device** (LRAD) and **Magnetic Acoustic Device** (MAD), which use acoustic energy.

Group

The fourth separation is by groups inside the categories. Groups within categories are differentiated by how a particular nonlethal option functions. For example, groups within the mechanical category would separate options that work as barriers, entanglements, or restraints of some sort. Likewise, groups in the impact category could be separated into handheld batons, launched projectiles, or multiple pellets flung from an explosion, as in a stingball grenade. Groups in the chemical category could be separated by whether the agent is primarily an irritant, malodorant, calmative, **obscurant**, and so forth.

Features

The fifth and final section is by dividing the various groups into their intrinsic features. This division separates the various options by the different characteristics of the device itself. For example, projectiles may take the form of **non-stabilized**, **spin-stabilized**, **fin-stabilized**, **drag-stabilized**, or **ring airfoils**.[9] Chemicals may take the form of powders, liquids, aerosols, vapors, and so forth.

Although this taxonomic system is far from complete, it is my best attempt at a précis of the collective knowledge of some of the world's experts on nonlethal force. To be sure, there are unresolved issues and differences of opinion and they need refinement with discussion and collaboration to bring more clarity and accord. In fact, about all that can be said definitively about this subject is that there is a definite need for common terminologies, definitions, descriptions, and understandings.

The remainder of this book will focus on the various categories of nonlethal force options.

Part II

Nonlethal Force Categories

10

Nonlethal Impact Weapons

IMPACT WEAPONS HAVE long been the bread and butter of nonlethal force. They are called impact weapons because they forcefully strike some object, usually a body, to create an effect. Because motion is required, they are also referred to as kinetic weapons.[1] Currently, and for the foreseeable future, every option in this category works on pain compliance; that is, they rely on inflicting enough pain on a belligerent to diminish his[a] will to resist. Pain compliance has proven an effective method of achieving tactical objectives, and to be perfectly frank, we in law enforcement are quite comfortable with it. It's the trauma that is used to achieve it that is problematic.

The predominant difficulty is that the nerve endings that transmit pain are clustered in areas of the body that are also more vulnerable to serious injury, especially the hands, face, and groin. The damage can be reduced by avoiding these areas, but it can never be eliminated.

In fact, in every street fight I've been involved in, the suspect has always been less than cooperative with my benevolent attempts to strike him safely[b] and more than one went to the hospital for stitches in places that made reporting problematic with my supervisors.

Notwithstanding my best efforts, striking a combatant *anywhere* resulted in some trauma, often requiring medical treatment. And while

a Female subjects who require significant force to subdue represent a very small minority when compared with their male counterparts. Admittedly, "his" is a male descriptor, and while there is no intent to exclude females, I can't believe that even the most ardent feminist would object to using male pronouns for crooks.

b Sarcasm is always risky lest it be misinterpreted, but, in this case, it serves to highlight the point that there are those who actually believe street fights more closely resemble sports, like wrestling or boxing matches, instead of vicious dog fights.

I have boundless sympathy for the aches and pains suffered by a suspect actively resisting arrest, even more troublesome is that this nonlethal method is also among the highest for injuries to the good guys (and I have the scars to prove it).

This is because the nature of these short-range weapons requires us to close with our assailant, making us vulnerable to kicks, blows, and thrown objects. Further, pain is highly subjective. Pain that would be devastating to one individual may be only mildly irritating to another, especially if they are under the influence of some drugs, are highly emotional, or are mentally distraught.

The category of impact weapons is generally divided into four groups distinguished by how they are employed. These are handheld, launched, thrown, and by projected energy. Although fists and feet can certainly be weapons, and I have personally investigated a number of murders and innumerable beatings with weapons no more sophisticated than these, for our purposes here I will stick with the more conventional understanding of weapons as tools and instruments.

Handheld Impact Weapons

The first group, handheld impact weapons, is comprised nearly entirely of clubs or whips of one sort or another. Clubs are a common tool in every police department on the globe. To be sure, we've provided them with euphemistic names—like truncheon, riot stick, baton, nightstick, and so forth—but they are, after all, simply clubs.[2] Although their originating purpose was certainly intended as a lethal weapon,[3] they have been used as nonlethal options for hundreds of years.[4]

The fact that they are not only still in use but routinely employed by law enforcement agencies throughout the world sends two irrefutable messages. The first is that they work. Nothing remains in service that long without providing some utility. The second is that nothing better has come along to replace them. This speaks volumes for the relatively primitive nature of the nonlethal weapons field as a whole.

Whips[5] are another ancient impact weapon, but rather than being euphemistically labeled, their names—like the **sjambok**, **bullwhip**, or **cat-o'-nine-tails**[6]—describe their particular attributes, and work by inflicting a stinging lash rather than the blunt trauma more often associated with a club.

Launched Impact Munitions

The second group is comprised of impact weapons that are launched. They are usually just called impact munitions, because many are fired from otherwise lethal launchers, most commonly the 12-gauge shotgun and 37mm or 40mm grenade launcher.[7] This group is the largest and most diverse of all nonlethal options and has commonly been labeled rubber bullets. Notwithstanding their name, they are more often made of plastic, wood, foam, lead shot, sand, and other materials. They are a more modern attempt to use motion and impact to inflict pain, but at ranges that provide a safer standoff distance. These munitions come in a variety of forms and shapes, like pellets, cylinders, round-nosed, and even bags.

One of the earliest versions is that of wooden dowels fired at the ground in front of an adversary, which then ricochet up into the legs of a belligerent, thus their nickname, knee knockers.[8] A more modern version is called a stinger cartridge, which fires pellets like rubber buckshot and is just as often direct-fired as **skip-fired**. One of the latest versions is sponge grenades,[9] which are intended for direct-fire and consist of a soft sponge-like material. The most modern versions are reliably able to strike a belligerent at a range of fifty yards.

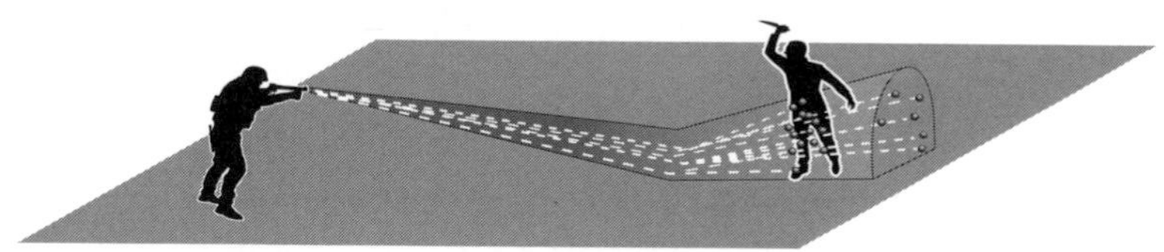

Figure 9: Skip-Firing

One way of reducing the dangers associated with firing multiple projectiles is by "skip-firing." Sidewalks and pavements are highly reflective to impact munitions, especially pellets, baton rounds, and the like. By striking the surface slightly in front of a subject, the beaten zone is halved while the saturation is doubled. Thus, more energy is transferred to the intended subject with less likelihood of striking an unintended target.

Thrown Munitions

The third group includes those nonlethal weapons that are thrown. They are distinguished from those that are handheld because they function after leaving the control of the user. The most common type is called a stingball, which looks like a black rubber softball, but explodes and flings small, hard rubber pellets that sting when they hit, hence the name. One of the advantages of stingballs is that they provide an ability of striking targets in **defilade**—that is, combatants who are using obstacles to shield them from projectiles and the like. This situation commonly occurs during riots, when agitators and provocateurs throw missiles at law enforcement officers or peacekeeping forces from behind cars, buildings, and even other members of a mob.

Stingballs come in a variety of configurations, most commonly with different-sized pellets ranging from about .20 caliber to well over .50 caliber. The advantage of the smaller pellets is that they provide a much greater saturation but, because they have so little mass, the effective radius is greatly diminished. The larger pellets provide a much wider radius but, because there are fewer of them, there is a greater likelihood of missing some of the agitators altogether. In addition to the size of the pellets, some stingballs include a dust of CN, CS, or OC powder that works as an instant, but relatively small and short-lived, tear gas cloud. Whereas stingballs can be thrown to about forty yards, modern stingballs can now be launched from a shotgun launcher out to about a hundred yards.[10]

Projected Energy Weapons

The fourth group is projected energy.[11] Admittedly, for the time being at least, this group is empty. In fact, there is only one promising candidate to fill it in the foreseeable future; that is the pulsed energy projectile (PEP). Using a chemical laser, it fires an invisible infrared pulse at a target. Upon contact with the target a small amount of plasma[12] explodes, resulting in a sound and shock wave, as well as some electromagnetic radiation, that can create a painful burning sensation. Unclassified reports suggest that ranges will be effective to well over one mile. Because it travels at the speed of light and is in line of sight, lag time, environmental factors, and cross-contamination are eliminated. Because it exceeds the ranges of many handheld lethal weapons, it will be capable of employing nonlethal force to prevent or deter lethal attacks. Reportedly in the late stages of development, there is no current estimate of when it might be completed. Moreover, because impact of any type is problematic in and of itself, the human bioeffects testing may delay an actual deployment for years afterwards.[13]

Challenges

The ultimate nonlethal force option will not rely on impact. There are simply too many biological and physics issues. For example, nearly all impact munitions are projectiles of some sort and are plagued by a physics problem called trajectory degradation. All this means is that once a projectile leaves the barrel of its launcher and is no longer being propelled, it is falling. Accordingly, the amount of energy attached to a projectile diminishes with distance and the projectile eventually falls to the ground. Whereas this is true of all projectiles, it is particularly troublesome with nonlethal munitions because of their slower velocity and greater weight.

Since the idea is to transfer just enough energy to an adversary to cause sufficient pain to overcome resistance but not so much as to cause serious injury, it presents some complex physics problems. The first is that designing a munition that provides an effective amount of energy transfer[c] at close ranges means that there will not be enough energy to work at long ranges. Conversely, a munition designed for greater standoff distances, especially those most often encountered in military applications, may make it permanently disabling or even lethal at closer ranges. Thus, every one of these munitions has a "sweet spot" that is beyond the minimum safe distance but short of the maximum effective range. (See Figure 10 on the next page.) The larger the sweet spot, the more "forgiving" the round. The fact that each munition has a slightly different sweet spot is bad enough, but the problem is complicated still further because, even in the best of circumstances, tactical situations seldom lend themselves to precise range estimation[14] and target acquisition.

---

c This energy transfer is just as often described as an energy dump.

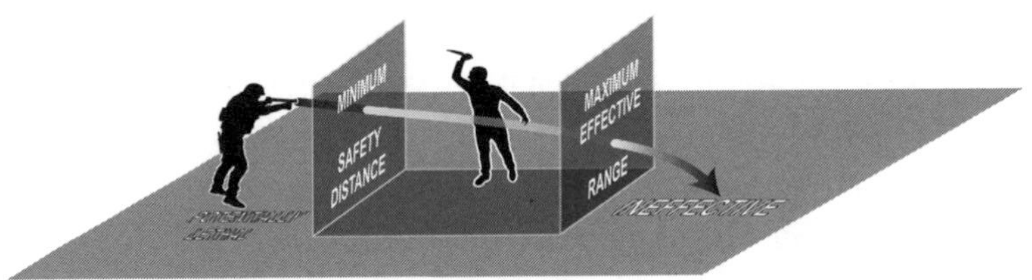

Figure 10: Impact Munitions "Sweet Spot"

Because nonlethal projectiles can be deadly at close ranges and ineffective at longer ones, a "sweet spot" exists, which incorporates both safety and effectiveness considerations. The larger the "sweet spot," the more forgiving the round. While simple in concept and easily duplicated in laboratory settings, it is extremely complex in application and extremely difficult to duplicate in the field because suspects do not cooperate, and ranges can change dramatically in the blink of an eye.

Besides the physical characteristics of impact projectiles, there are biological issues as well. Where an adversary is struck is just as critical, if not more so, than what he is hit with. When it comes to sustaining impact, the human body is not of consistent resiliency. The muscular parts of the body—like the sides of the back, thighs, and arms—shield the skeletal bones underneath, while the most vulnerable organs—like the spleen, heart, lungs, kidneys, liver, and so forth—are safeguarded by a rigid rib cage that is amazingly capable of protecting them. When it fails, however, it fails catastrophically, and most deaths from impact munitions have been when a projectile penetrates the chest or breaks a rib that lacerates a vital organ.[15]

Because most police agencies, and nearly all military organizations, have been using "center mass" as the aiming point, the likelihood of hitting the chest area is increased. (See Figure 3 on p. 42.) As a result, many of the more progressive agencies have determined that a safer aiming point is the belt buckle. While recognizing that an injury to the groin or abdomen is likely, these injuries are usually not immediately life-threatening and will allow time for medical attention, unlike a chest injury, which is more likely to result in immediate death.[16] Although some tactical situations may even allow targeting the extremities, it is generally impractical given that these are small targets and nearly always moving.

There have been many attempts, and even some recommendations, to identify the amount of energy necessary to be effective but remain short of being seriously injurious.[17] The transfer of energy, however, is a factor of both design and velocity. When we consider that a young college student in Boston, Massachusetts was accidentally killed after being struck in the eye with a projectile that imparted only about twenty-five foot-pounds of energy, but that suspects routinely ignore the impacts of 37 and 40mm baton rounds[18] that consistently release in excess of ninety foot-pounds of energy, it is clear that there is no completely reliable fail-safe formula. In fact, all formulas and algorithms attribute way too much precision for situations that are inherently unpredictable and chaotic. To be sure the math is there and is expressed as $E = \frac{1}{2}MV^2$. Translated, this just means that the amount of energy is equal to one-half the mass (of the projectile) times the velocity squared. Changing the speed of a projectile has a far more dramatic effect on the amount of energy than changing the weight.[19]

That is only the first part of the equation, however, since the surface area over which the energy is transferred is just as critical. In America, energy is measured in foot-pounds.[20] A foot-pound is simply the amount of energy released from a one-pound object dropped from one foot. As you've certainly figured out by now, I'm no rocket scientist, so I've had this issue explained to me by several mechanical engineers and the best way I've understood it is with the following example.

Imagine a metal plate that is one foot square and one inch thick and that weighs exactly one pound; dropping it from one foot will release exactly one foot-pound of energy. If that same plate were cut into one-inch square cubes and stacked into a vertical column, however, the weight would still be one pound and would release the same amount of energy, but the energy would be transferred into a vastly smaller area. This is precisely the problem with many of the commercially available impact munitions that advertise minimal amounts of energy but fail to consider the size and shape of the projectile as it strikes a human body.[21] It is the same reason a pencil lead will easily push a hole through a piece of paper but you can turn it over and press hard to use the eraser.

Standards of Accuracy

Another of the more difficult problems with this category of nonlethal options is accuracy. As the years have passed and nonlethal impact munitions have become abundant, three different standards have emerged. Initially, attempts were made to require the same degree of precision as that required of lethal options; that is, point of aim equals point of impact. To date, only at close ranges can nonlethal projectiles come even close to this standard, however, and so a new standard evolved.

The second standard is a percentile score of hitting a man-sized target at a given range. For example, "This projectile will strike a man-sized target 80 percent of the time at forty yards." This standard still exists but has fallen from favor in law enforcement circles. To illustrate why, consider the following actual case that occurred in Los Angeles in late 2000.

A patrol car encountered a vehicle reported as stolen. After the vehicle was stopped, the occupants were ordered to exit and kneel as per procedures for a high-risk encounter. One of the passengers was a woman who made some movements perceived as a threat by the officers, one of whom was armed with a shotgun and stun bag munitions (also commonly known as bean bag munitions). As a result, he fired three times. One of the projectiles missed the woman entirely, one struck her in the back, and one struck her in the right eye.[22] According to the second standard of accuracy, both the developer and law enforcement would consider the first round a miss and both would consider the second round a hit. A dispute arises with the third round, however, since it meets the definition of striking a man-sized target for the manufacturer, but law enforcement would consider it a miss in that it did not go where the officer intended. Marksmanship was never a serious issue in that any officer can be expected to hit a stationary target three times out of three from a few feet away; therefore, the problem can be clearly attributed to the inconsistency of the munition.

This incident, and others like it, have been most upsetting to the public and law enforcement alike and resulted in a proposed new standard as a result of a study done by Penn State's Applied Research Laboratory.[23] Recognizing that, for the foreseeable future at least, nonlethal munitions are unlikely to achieve the same degree of precision as lethal munitions, the alternative standard of accuracy was also unacceptable. Consequently, the Penn State researchers proposed a new standard, which would simply measure the smallest diameter circle to include a given number of impacts[d] at a given range. Obviously, the smaller the diameter, the more consistent the munition and accurate shot placement could be achieved by simply adjusting the aiming point of the launcher. This standard allows for the greater effects of trajectory degradation for nonlethal munitions without compromising the expected results of the shooter.

Projectile Composition

And if you thought that range, accuracy, and energy were not enough to complicate the development of a safe and reliable nonlethal projectile, what it is made of also seems to influence the degree of anticipated injury. Nonlethal projectiles are made from a variety of materials, including wood, rubber, foam, plastic, lead, and even sand. Some are even made of soft rubber and filled with a liquid. When dealing with a single composition, such as wood, rubber, foam, or plastic, hardness is measured by using a **durometer** scale.[24] A measurement of 100 is about as hard as a piece of wood, while a reading of 10 to 20 is about the resistance of a dry sponge. Human skin varies in hardness,[e] but generally is about 20. Accordingly, material that has a rating of 20 or less will be less likely to penetrate a human body because it tends to "flare" and spread the energy over a larger surface.

Logically, the harder the material from which a projectile is comprised, the more likely it is to result in lacerations and penetrations when it hits human skin even a glancing blow. In fact, however, every projectile has the potential to cause greater injury if it strikes any portion of the body where bones are close to the surface of the skin. Areas like the shins, elbows, collarbones, and especially the head, have only a few millimeters of tissue between the hard bone and the surface. When struck by a projectile, even one that seems soft, the tissue is compressed between

---

d In the case of this study, it was five.

e As was noted on p. 42, it also varies in thickness, from about .5mm on the eyelids to as much as 4mm on the palms of the hands and soles of the feet.

the bone and the projectile, often resulting in severe lacerations and gouging. While softening a projectile has some effect, avoiding these areas of the body entirely has proven to be a better solution.

Conclusion

Despite their problems, impact nonlethal weapons have been the mainstay in the array of nonlethal force options, probably because they are cheap to employ and can provide a respectable standoff distance. Even so, they are debilitating and not incapacitating. Pain is highly subjective, and any force option that relies solely on the amount of pain it can inflict is inherently unpredictable. Notwithstanding, even primitive weapons and munitions that fall short of the consequences of their lethal counterparts will prove beneficial. To reiterate, when it comes to nonlethal force options, the standard is not perfection; the standard is the alternative.

11

Chemical Agents

NONLETHAL CHEMICAL WEAPONS are those that rely on some pharmaceutical interaction with the body. This is an important distinction in that other nonlethal options use chemicals but do not rely on this interaction, and so would not be classified as chemical options. Two classic examples are sticky foam and anti-traction technologies. Both are chemicals, but neither requires a pharmaceutical interaction. Sticky foam works by physically impeding a person's mechanical ability by retarding (gluing) limbs to objects, including other limbs. Thus, sticky foam more closely approximates the operation of a **net**. Similarly, anti-traction technologies (super slippery) work by reducing the friction necessary to easily function, so much so that even standing becomes difficult. Accordingly, these options more closely approximate those assigned as nonlethal mechanical options. A true nonlethal chemical option, however, works by interacting with the organs of the human body to interfere with vision, balance, alertness, breathing, sneezing, coughing, and so forth.

Background

Historically, chemical agents are preceded only by impact weapons as nonlethal options. The use by the Chinese of finely ground lime dust as a riot control agent to quell a peasant riot in 178 C.E. would certainly meet the modern definition of a nonlethal option. They dispersed the dust into the wind with horse-drawn chariots equipped with bellows.[1] It wasn't for another thousand years, however, that another example meeting the modern definition occurred when the Japanese used the Metsubishi to disperse ground pepper. The Metsubishi is a small round or square box, or sometimes an eggshell, which could be filled with a powder. It was reportedly used by Japanese law enforcement officers to disable an adversary for capture.[2] The pepper dust was dispersed by either blowing through a mouthpiece or throwing the eggshell, which burst when it hit an adversary, releasing a cloud of pepper. Needless to say, given the span of centuries in between, neither of these methods was a trend setter and it was not until the Parisian police used tear gas[3] to capture a gang of bank robbers[4] in 1912 that it could be said that the concept of nonlethal force as an objective had caught on.

Employing Chemical Options

Chemical options can take any of the three forms of matter, solid, liquid, or gas. Many tear "gases" for instance, are simply solids that are powdered so fine that they are lofted and suspended by hot gases. Others are dispersed as a liquid in an aerosol dispenser. A true gas is far more difficult to effectively deploy and is reserved for the more exotic nonlethal chemicals, like calmatives and hypnotics.

In order for a chemical to have a pharmaceutical reaction it must enter the human body in some manner. There are only four ways this can be achieved: inhalation, absorption, ingestion, and injection. Inhalation is commonly used for riot control agents because of the ease of employment and ability to simultaneously affect large numbers of people. Although absorption through skin is difficult to employ because of clothing, it is commonly used with liquid aerosols by spraying directly on the face, especially the eyes. Arguably, ingestion is the most difficult of all, because it requires an adversary to eat or drink the chemical. Injection is a method commonly used against animals, such as with tranquilizer darts, but is nearly unheard of against humans.

Tear Gases

Nonlethal chemical agents are grouped by how they function. Currently, there are three groups in the chemical agent category: tear gases, malodorants, and calmatives.[5] The first group in the chemical agent category is tear gas. The term "tear gas" was chosen over "riot control agent" (RCA) because the same chemicals are just as often used in other situations. Like "rubber bullets," the term "tear gas" has become a generic term for all irritant chemical agents, often resulting in tearing, coughing, sneezing, irritation, and even pain.[6] Although there are noted differences between the agents, especially OC, CS and CN, they are all used in the same manner and for the same purposes; hence it makes sense, while noting their differences, to keep them in the same group.

The early success by the French police, coupled with the public's increasing demand for less violent methods for handling mobs and riots, provided an impetus for more improved chemical agents. As a result, Federal Laboratories was formed in 1923 to provide chemical munitions to American law enforcement agencies. Of the tear gases used in the United States, CN is the oldest and was in use from the 1920s through the early 1960s. CN stands for chloroacetophenone. It was invented by a German chemist in 1869, who noted its ability to cause tearing of the eyes.[7] In low concentrations it has an odor of apple blossoms and is very quick reacting, usually causing heavy tearing and visual impairment in only a few seconds. Besides causing the eyes to tear, a burning, itching, runny nose is also common. In heavy concentrations it may even blister the skin, and although rare, deaths have occurred. CN fell from favor with the advent of CS, which proved far safer and more effective. CN was banned by the US military in 1964[8]; however, it remains commercially available. CN munitions can be identified by red markings on grenade canisters and projectiles.

Unlike CN, which is a moniker for the chemical name, CS was coined from the initials of the British chemists who developed it. B. B. Corson and R. W. Stoughton developed 2-chlorobenzalmalononitrile[a] in 1928 and noted that it worked very similarly to CN. In low concentrations it has a distinct peppery odor. Although it is not as fast acting as CN, it also causes tearing eyes and an itchy, watery nose, and could cause a

---

a The pronunciation of this polysyllable word is reason enough to find a substitute. CS has been classified as both a lachrymator and an irritant.

burning sensation to the skin if the latter is exposed to it for more than about ten minutes. CS also causes a burning sensation in the lungs and a feeling of tightness in the chest, sometimes resulting in feelings of panic from the perception of being unable to breathe. The primary reason that CS replaced CN as the preferred tear gas, however, was that although the effects are more severe, it is far safer.[9] CS munitions can be identified by blue markings on grenade canisters and projectiles.

Currently, the most popular nonlethal chemical agent is pepper spray, so called because it is derived from peppers and is normally applied by spraying a belligerent in the face, especially the eyes. It is just as often called OC, which stands for the active ingredient, oleoresin capsicum. "Oleoresin capsicum" is a horticultural term that is an ingredient in chili peppers. In fact, the OC in nearly all commercially available dispensers is a derivative of Cayenne, Jalapeño, or sometimes Habañero[b] peppers. These contain the alkaloid, capsaicin, which, in its purist form, is tasteless and odorless but can be detected by humans at one part in ten million!

Pepper spray has been in use as a deterrent against animals, especially dogs, since at least the mid-1970s. It gained recognition when the US Postal Service began issuing it to mail carriers as a deterrent against vicious dogs, who are immune to the effects of CN and CS. It was introduced to American law enforcement in 1982 at the annual International Association of Chiefs of Police Conference, and the FBI began testing it in 1987 and approved it for use by their agents later the same year. Within a few years, it had gained widespread acceptance throughout American law enforcement and is one of the most popular nonlethal force options in American law enforcement.[10]

The effects of OC are the immediate, involuntary closing of the eyes, accompanied by a burning sensation on the skin and a feeling of difficulty breathing. Although it is sometimes grouped with the other tear gases, it is usually classified as an inflammatory. Moreover, it is seldom used in support of riot suppression. This is because, whereas CN and CS have a decontamination time of about fifteen to twenty minutes, the

---

b Habañero peppers are generally considered to be some of the spiciest peppers in the world.

debilitating effects of OC easily last as long as forty-five minutes,[11] with residual discomfort lasting several hours, creating an increased danger to those exposed. Likewise, it is heavy and quickly falls to the ground, which makes contaminating large areas more difficult. Besides the advantage of its immediate effects, OC has another advantage over CN and CS. It works on animals, including dogs. In fact, long before it was in use by police departments, postal workers were carrying it, for the obvious reasons.[12] OC munitions have not yet been assigned any universally accepted color code, but orange is the most prevalent.

Although there are other nonlethal chemical agents exist for antipersonnel and riot control, they are not generally available or widely used. Two that merit mention are **CR** and **DM**.[c] CR is a pale-yellow crystalline substance with a pepper-like odor. It was first synthesized in 1962 by British scientists R. Higginbottom and H. Suschitzky. It has the advantage of being more potent but less toxic than CS and was manufactured by the Ministry of Defence in the United Kingdom from 1968 through 1977. It was authorized for use in Northern Ireland in 1973 and approved by the US Army in 1974,[13] but never gained wide acceptance due to a lack of supporting scientific testing.

DM was developed in 1913 by American chemist, Roger Adams, of the US Chemical Warfare Division.[14] DM is also referred to as **adamsite**[d] or **vomiting gas**. It is a yellow solid with almost no odor. As you probably guessed, DM causes vomiting, as well as nausea, coughing, and weakness. Although the effects of DM are far more severe than other tear gases, it is notoriously slow acting, taking as long as six minutes before becoming effective. Consequently, DM was often mixed with CN in order to cause some reaction immediately. DM was used in Washington D.C. on July 28, 1932 to break up a protest.[15] Although neither of these agents is currently being used, they serve as examples of successful employments of nonlethal chemical agents, as well as providing guidance for acceptability and suitability for future ventures.

---

c In seeking to spare you the burden of reading the chemical names, I omitted them. For those sickly curious creatures who really want to know, CR stands for dibenzoxazepine and DM stands for diphenylaminochlorarsine.

d For the obvious reason.

Malodorants

The second group is reserved for malodorants. Malodorants take their name from their description; that is "mal," meaning bad, and odor. In short, they reek! Actually, it is difficult to find adequately descriptive terms to describe just how badly they stink, but I can assure you they'll gag a maggot! I once gave a demonstration in a large field near Brownsville, Texas attended by three hundred or so local and federal police officials. After carefully checking the wind, I opened the top of a small glass vial with an opening of approximately 3/16ths of an inch containing about 0.5cc of a malodorant being developed by a company in Texas. Holding the vial about shoulder height while I talked, the wind blew slightly from behind me as I watched (admittedly with sick delight) while people pushed and shoved to move out of the airstream. Within a very short time the audience was neatly divided into two groups with a cone-shaped split of stench between them. Nothing I could have said would have made the point any clearer.

Malodorants are just beginning to be seriously considered[16] as a nonlethal alternative. Nonetheless, they promise at least five advantages over conventional riot control agents. The first is that malodorants are effective in extremely small amounts, often less than one part per billion. Accordingly, the logistical burden[e] in storing, transporting, carrying, wielding, and so forth is reduced dramatically. The second is that they are self-dispersing. All this means is that once a provocateur is contaminated with a malodorant the substance will continue to be dispersed as it blows off his clothing and body. This creates somewhat of a dilemma for him. If the individual moves toward the peacekeeping force, he will be arrested, but if he moves back into the mob he carries and releases the malodorant, effectively dispersing the crowd as he moves through it because no one can stand to be around him. Moreover, anyone who also becomes contaminated, as in brushing against him, also becomes a dispersing agent.

In this manner he becomes the functional nonlethal equivalent of Typhoid Mary.[17] Third, law enforcement and peacekeeping forces are

---

e This is usually a bigger problem for the military than law enforcement.

Figure 11: Malodorants Are Self-Dispersing

often confronted with a tactical dilemma of their own when a provocateur is inciting to riot.[f] If he is allowed to continue, the chances of a riot are more likely, but if authorities attempt to arrest him they run the danger of starting the very riot they are trying to avoid. They will afterwards be blamed for inaction if the riot starts without an intervention and overreaction if it starts because of their attempts to remove the provocateur. Dousing him with a malodorant, however, will quickly remove his audience without requiring an arrest.[18] Fourth, malodorants can be formulated for different durations—from minutes to weeks. As an anti-personnel option, the effects can be tailored more precisely than conventional riot control agents, and as an area-denial capability, a malodorant can provide some safeguards for empty and unprotected buildings or adversary staging areas during riots or other disturbances. Fifth, counteragents that eliminate the stench of the malodorant have been developed and have proven effective. Consequently, a small amount applied underneath the nose will eliminate the need for a gas mask.

f Inciting to riot is a crime in just about every state in the union, including the federal government, as well as in most countries.

Malodorants are nearly always synthetic and use a variety of ingredients, to include a compound causing the smell with a liquid carrier. Typical ingredients include organic sulfur; skatole, which smells like fecal matter; mercaptan,[19] with a smell like rotting cabbage; and butyric acid, which smells like vomit. Other compounds smell like rotten eggs, vomit, dead fish, skunks, or putrid garbage. Depending on the type of contaminant and concentration, effects range from mild revulsion to gagging and vomiting. Malodorants are generally launched rather than sprayed, since any overspray or blowback contaminates the user. One of the most successful methods is by encapsulating the malodorant in a projectile that bursts on impact. Others include small, brittle globules that can be thrown or launched.[20]

Calmatives

The third group is calmatives. In the simplest terms, a calmative is a drug that has a sedative effect, which lowers functional activity. Calmatives work by reducing an adversary's ability to have an effective response, even by putting them to sleep. Like the grouping of tear gasses, the calmative group includes agents that have slightly different effects. Consequently, other suggested names for this grouping have included somnolents, soporifics, hypnotics, and narcoleptics—all of which are variations of drugs that cause drowsiness or sleep. Their common characteristic, however, is that they produce a tranquilizing effect of torpidity or unconsciousness.

Calmatives provide tactical advantages that cannot be attained by any other nonlethal option, not the least of which is separating victims from combatants. The most notable example to date occurred in October of 2002 at the Dubrovka Theater in the southern outskirts of Moscow, Russia when approximately fifty armed and suicidal Chechen rebels took an audience of nearly 900 people hostage and demanded an end to the Chechen War and the withdrawal of all Russian forces. After several days of largely futile negotiations, a calmative[21] was pumped into the building, rendering nearly everyone unconscious. Although more than 700 hostages were rescued, as many as 170 died as a result of the gas, resulting in a storm of controversy.

Ideally, calmatives would provide a nearly perfect solution for hostage situations, in that putting everyone to sleep would allow authorities to safely separate the victims from the criminals, providing an ability to rescue the victims and arrest the crooks without violence. It would be especially advantageous when other tactical means are not viable, as in airplanes, schools, theaters, and the like.

Less conspicuously, and on a far smaller scale, similar situations occur frequently throughout the world. This has resulted in a demand for better calmatives, more efficient methods of introducing them, and effective responses to include counteragents. Countervailing opinions insist that calmatives are too dangerous even to explore.

An ideal nonlethal calmative agent needs to meet at least four criteria. First, it needs to be easy to administer and adaptable to any of the four methods necessary to introduce any chemical to the body. Furthermore, chemical agents that require extraordinary storage conditions (like temperature, humidity, pressurization, and so forth), or are difficult to transport and handle, pose more difficult tactical challenges. Likewise, nonlethal chemical agents need to be adaptable to circumstances because circumstances will not adapt to accommodate them. In the words of one of my partners, tactical responses are "come-as-you-are parties"[22] and a calmative that is difficult to administer quickly becomes a pivot point for planning.

Second, the onset of the effects must be fast acting and of short duration. With few exceptions, the effects should take only a few seconds and the duration should not last more than a few minutes. Slower acting calmatives provide opportunities for detection and countermeasures and longer durations become dangerous to health. Third, a given dose should produce a predictable and reliable effect to the widest range of people possible. This means that the same dose will be effective on a large, healthy person without endangering smaller people or those in poor health. Fourth, the effects should be quickly reversible, either by being metabolized or by the introduction of a counteragent, without lingering effects. Calmatives that have even minimal residual effects, such as dizziness or nausea, complicate evacuations and rescues.[23]

The most critical problem in using calmatives is the same one that plagues all chemical agents: the **dosage factor**. In the simplest terms, the dosage factor just means that the amount of agent necessary to be effective against one person may be harmful or fatal to another, and one that is safe for all may be completely ineffective for some. Whereas the dosage factor is applicable to all chemicals, calmatives are the least forgiving.[24] An overdose of tear gas is not only hard to achieve, the onset of symptoms are easily noticeable by even the untrained, not to mention the person exposed, and efforts to avoid additional discomfort are nearly automatic. With calmatives, however, the effects may not only be unnoticeable, but an overdose can be fatal.

Whether used alone or to enhance the effects of other nonlethal options, calmatives will remain controversial long after safe and effective agents become available. Nevertheless, incidents like the Dubrovka Theater in Moscow, the Branch Davidian complex in Waco, Texas and scores of other lesser known but just as dangerous incidents, require better nonlethal capabilities than currently exist. Calmatives are one of the few nonlethal options that show promise in effectively handling situations that have historically required lethal force.

12

Mechanical Options

NONLETHAL MECHANICAL OPTIONS are those that restrain the intended motion of something. One of the critical components of this definition is the word *intended,* since some mechanical options work by reducing friction to the point where motion is greatly facilitated, just not in the intended manner. The mechanical category is sparsely populated when compared with either chemical agents or the impact categories, especially with options that are primarily intended as anti-personnel. Nearly without exception, mechanical options involve some type of entanglement that physically restrains motion of some type. A case might be made for handcuffs and other restraining devices as nonlethal options; however, I have omitted them because they fall outside the commonly accepted understanding of force.

Mechanical options come in a variety of configurations, but most commonly constructed as nets or **running gear entanglement systems**. At risk of stating the obvious, nets are simply a woven, meshed fabric to catch or ensnare a person or other object. Running gear entanglement systems are made from ropes or cables and designed to foul propellers or entangle the wheels and powertrain of a vehicle. Two other configurations use chemical substances.[1] Similar to glue, sticky foam is extremely tacky and adheres to just about anything in contact, to include clothing and human skin. The other is the diametrical opposite and works by making things too slippery to grasp or even remain erect! Called **super slime**, it is a powder that, when combined with water, becomes a viscous goo that is as slippery as wet ice.

As anti-personnel alternatives, nonlethal mechanical options have had limited success. From my earliest days as a rookie on the streets of South Central Los Angeles when "angel dust"[a] was prevalent, I have seen nonlethal mechanical options such as long poles with a chain or rope hooked between them to ensnare a belligerent's feet; ladders used to pin a combatant against another object; wet blankets thrown over a belligerent to drag him to the ground; as well as many others. To say that these were less than effective gives them far more credit than they deserve but speaks volumes for the desperation we were experiencing in handling people who had no perceptive pain threshold.[2]

Capture Nets

Nets have been successfully used for decades to capture fish for eating and wildlife for research so it would seem they would have equal utility as a nonlethal option against humans. They have also captured the imagination of well-wishers and seem appealing as an "easy fix" for capturing combatants without causing severe injury. In practice, however, they have not lived up to these expectations.

It is unlikely that nets, in any form, will ever be more than a novelty in any anti-personnel role for nonlethal force options for the following reasons:

First, nets are difficult to employ and, whether launched or thrown, move more like a sail than a projectile. Besides not being aerodynamic, they are relatively slow and short-ranged. An adversary, especially one that is moving, can evade them without much difficulty.

Second, they are nondiscriminatory, meaning that it is impossible to target a combatant in proximity with another person without ensnaring them both.

Third, they use weights to unfold the net and each of these weights is an unaimed projectile in and of itself. Even under ideal conditions, role players demonstrating how they might work wear helmets for protection.[3]

Fourth, when a person is inside a building, the net often strikes a wall or ceiling with enough force to bounce away without entangling the

a Also known as PCP (phencyclidine hydrochloride), and a variety of street names.

suspect. Likewise, if they are near bushes or shrubbery, the net becomes caught in the branches and leaves without seriously impairing the intended target. This makes them impractical inside all dwellings and many other buildings.

Fifth, even if the net succeeds in ensnaring the intended target, it can take a considerable effort to safely remove the net before a suspect can be handcuffed and transported.

Sixth, an agitated suspect who is waving his arms or has a stick or something of the kind can foul the net before it ensnares them, rendering it completely useless. Because this behavior is not uncommon and is an effective countermeasure, it is likely that it will be quickly adopted by adversaries encountering nets as a nonlethal force option.

Suggestions that nets be electrified or use a glue-like substance to make them sticky have been considered and some even developed as a prototype. Notwithstanding the best efforts of the US Department of Justice, which has spent considerable time, money, and effort in trying to develop nets as an anti-personnel nonlethal force option, the inherent problems appear insurmountable and other options are cheaper and more effective.

Running Gear Entanglement Systems

Running gear entanglement systems (RGES) are nonlethal options in an anti-mobility role. The problems inherent in an anti-personnel role are negligible or do not exist when used in an anti-mobility role. They vary in configuration, but all employ some sort of a rapidly deployable rope or net-like system that can stop a vehicle or boat by entangling the running gear and/or drivetrain. Those designed for vehicles are sometimes anchored to a firm object and incorporate spikes to deflate tires, whereas those for boats typically use floats and loops to keep the system near the surface and to enable easy retrieval.

One of the most mature of the running gear entanglement systems was developed by the US Department of Defense and can stop wheeled vehicles up to 7,500 pounds moving at 45 mph. Called the Portable Vehicle Arresting Barrier, it is designed to be placed at checkpoints (such as gates or wharves), and chokepoints (such as bridges, tunnels, and other narrow passageways) to prevent passage of unauthorized vehicles. Other types of mechanical devices employed for similar situations include steel nets, retractable bollards, and anti-ram barriers.

Although not a running gear entanglement system, per se, **spike strips** are a variation that is popular with American law enforcement agencies. Spike strips work by using metal spikes, often hollow, to puncture the pneumatic tires of a vehicle driven over them. Some variations allow for the hollow spikes to detach and embed themselves in the tires to hasten the deflation. Although this certainly makes driving difficult, it does not always stop them, and numerous examples exist of vehicles continuing to be driven with all four tires flattened.

Similar in concept but different in application, are caltrop spikes, more often just referred to as caltrops. The name comes from the Latin *calcitrapa*, which means "foot trap." The idea originated from several plants that have spiked seeds that are extremely painful to step on. The spikes of the seeds are arranged so that they are nearly impossible to avoid, with one spike always pointed up. Manufactured versions have been traced back thousands of years and have been used since antiquity as protection against cavalry (including camels and war elephants) and infantry. These versions are typically iron but nearly all contemporary caltrops are made with steel and are especially effective against pneumatic vehicle tires. They have been used extensively in both world wars and are still used by extremists and militants. Although still used in military applications, there are no known uses by law enforcement agencies.

Vessel entanglement systems are also available and can be prepositioned or deployed via helicopters or other vessels to entangle engine propellers. These systems are quite commonly made from a series of loops along a polyethylene line, often with floats to make it more buoyant and effective by keeping it near the surface and to facilitate retrieval. The entanglement occurs as a vessel travels over the line and its propellers snag the line and become snarled.

A variation of entanglement devices uses a **bola system** as an anti-personnel option in which a strong but lightweight cord connects two or more weights that are sometimes padded. When launched or hurled against a fleeing adversary, the weights wrap around and entangle the legs or arms. One option, using small barbs as weights to enhance the entanglement, is now commercially available and is launched using a blank pistol cartridge. While this concept has proven effective against animals in South America for millennia, it has recently begun to show promise for use against humans.

Sticky Foam

Sticky foam is an expansive chemical that works as a human adhesive. It is categorized as a mechanical device because the predominant method of operation is not pharmaceutical but in restraining motion and is more appropriately placed in this category than the chemical category. It looks a lot like spray foam insulation but is unbelievably tacky. It adheres to nearly anything, including clothing and human skin. Originally developed as an area denial option by Sandia National Laboratories, it was conceived as a means for protecting sensitive facilities, like nuclear reactors and ammunition magazines.

The first time I saw sticky foam was on the airstrip in Mombasa, Kenya, while I was working with the Marine Corps during Operation United Shield in the early spring of 1995. My team[b] was comprised of some of the best experts in nonlethal options in the entire Marine Corps, and we would be training other Marines to enable a nonlethal capability for a mission that historically had required lethal force. All the training would be done afloat, primarily on the USS Ogden, as we prepared to go ashore into Mogadishu, Somalia to extract the United Nations peacekeeping troops. Besides providing training, my team's assignment was to develop and employ nonlethal methods and technologies to separate combatants from noncombatants. Though easy in concept, it was exceedingly difficult in application.

We were not only empowered but encouraged to explore all available options and one that attracted our attention was sticky foam. Although it

b Besides myself, the "tiger team" consisted of Science Advisor, Bob Walsh; USAF Lt. Rob Ireland; Marine Corps CWO-3 Jim Adams; and Gunnery Sergeants Tony Agurs, Pat McGilton, Mike Rodarte, T.J. Dunn, and Sergeant Bobby McCreight.

was developed primarily as a nonlethal area denial option, we thought it might have some use as an anti-personnel option and were given a launcher that would squirt it out to about twenty feet. We had been told by the scientists to aim it at the feet but quickly discovered that a suspect could move faster than we could "glue" his feet to the floor. I had pretty much given up on it when two of my gunnery sergeants brought me below decks to demonstrate a new technique they had developed.

Dressed in a disposable Tyvek suit, Gunny Rodarte, the "suspect," was armed with a large club, a behavior we anticipated encountering ashore. Gunny Dunn was playing the role of the Marine and when Rodarte tried to hit him with the club he squirted his crotch area rather than his feet. Instantly, Rodarte became a "penguin" when his thighs were glued together. Barely able to move and easily subdued, we realized that the previous failures were not due to the nature of the technology but how it was used. Not only was the feasibility of the sticky foam established, but the need to develop effective tactics, techniques, and procedures was unmistakable. Over the years, this lesson has been reinforced many times with other options as well.

The use of this specific formula of sticky foam never gained widespread acceptance, largely because of health concerns and the difficulties in clean up. Nevertheless, we demonstrated the utility of the concept and future formulas will undoubtedly address these problems. Its utility in protecting sensitive sites is self-evident and there have also been suggestions of using it as an anti-mobility option against vehicles.

Anti-Traction Technologies

Primarily used in a nonlethal area denial role, these technologies are also known as **mobility denial** systems. Rather than impeding movement, they work by interfering with an adversary's control over movement. In short, these technologies are slippery. Early prototypes have nicknames like snail snot, slippery slime, artificial ice, and instant banana peel. These are descriptive portrayals in that these options are usually a clear, syrupy gel that is slippery beyond belief! The latest formulas date to the 1990s; however, the effort can be traced back to the early 1950s. Modern versions, developed by Southwest Research Institute in San Antonio, Texas, can be applied as polymer-based granules that remain dormant until exposed to water.

When applied to nonporous materials, such as linoleum, tile, or wood floors, the gel is nearly impossible to walk on without falling. It takes little imagination to picture how difficult it is to climb stairs or to run! Other applications, such as over concrete, asphalt, brick, or wood, can be used outdoors. In fact, a portable system allows a single individual to carry and contaminate a 2,000-square-foot area, large enough to cover a typical urban intersection. Other possibilities include bridges, driveways, airstrips, and sidewalks. Besides pedestrians, wheeled vehicles are rendered virtually useless. Moreover, any incline enhances the effects. It even works on grass and gravel, albeit with somewhat diminished results.

What is more, the gel works best because it is hard to detect, especially in low light, and can be applied where it is least expected. Thus, an adversary must consider even untreated areas as suspect. It can also be used to coat objects to make them difficult, even impossible, to grasp or manipulate. Because it can also be sprayed on vertical surfaces, the gel works to make it difficult, or even impossible, to scale walls and fences, climb aboard ships, or grip ropes. Furthermore, it is easily transferred from one object to another and tends to adhere to the bottom of shoes and gloves, so there are residual effects that enhance the effects still further. Depending on weather conditions, the duration can last as long as twelve hours and new formulas make both longer and shorter durations possible. Likewise, it is environmentally benign and can be rendered ineffective with copious amounts of water, including rain, and will degrade over time.

13

Electrical Options

SOMETIMES CALLED ELECTRO-SHOCK devices, this category includes all the nonlethal options that rely on electrical stimulation for their effects. The term "electrical stimulation" is better understood as "electrical shock." An electrical shock occurs when an electrical current passes through the body of a human or animal. It results in a physiological effect that can be painful, although, for many electrical nonlethal force options, pain is a byproduct rather than the intent. Ironically, electrical options are some of the most studied and least understood of all nonlethal force options. In the United States, they are also among the most controversial.

Anyone who has ever been shocked can attest to the unpleasant experience, but without some knowledge of the factors involved, it is impossible to accurately assess the dangers. This is predominantly because of two reasons.

First, there are three influences that affect how electricity reacts with the human body. The amount of current (amperage) refers to the quantity, while the force (voltage) at which it flows can be understood as pressure, and the impedance (ohms) it encounters when in contact with a human or animal identifies the degree of resistance restraining the flow. These influences can be thought of as in a state of dynamic tension, in that each affects the other and changing any one also changes the other two.

Although simplistic, one way to understand how these influences interact with one another is to compare electricity with water. The amount of electrical current would be equivalent to the volume of water, with the electrical voltage the equivalent of water pressure. The electrical impedance would be like the restriction of water through a pipe: the smaller the pipe, the harder it is for water to flow through it in any given time.

Continuing with this analogy, the water pressure in your house is usually around 50 psi. This means that a typical garden hose[a] can deliver about seventeen gallons every minute, whereas a kitchen faucet will provide a little over two gallons of water a minute and a shower in the same house will deliver roughly the same amount. You can increase the volume of water by either increasing the diameter of the pipe or increasing the pressure pushing it through. Conversely, you can decide to use an aerator on the faucet and a low-flow showerhead to conserve water, but the pressure of the water will remain the same. This is because you have only reduced the amount of water available, not the pressure. So, it is with electricity. It is the amount of electricity (amperage) not the pressure (voltage) that increases the hazards. Consider that a static shock from walking across a carpet can generate as much as 35,000 volts, while a wall plug has only about 110 volts. Yet it is the wall plug that poses the greatest danger of injury.

Second, the conditions under which a person or animal is exposed to electricity, to include the part of the body in contact, has a great effect. Humans, for example, are usually covered by clothing, which is usually a very poor conductor of electricity and thus offers great resistance. Some clothing, such as rubber boots and gloves, are so resistive that electricians wear them as protection. When in contact with the human body, dry skin offers about 100,000 ohms of resistance; but just underneath it, the muscles and organs are as much as fifty times lower! Understandably, a break in the skin can have a major effect. Likewise, whether the skin is wet—whether the dampness is from rain or sweat, or even moisture in the atmosphere—can affect the severity of an electrical shock. Whether it is injurious or not depends on the amount of electricity, the duration of the exposure, the general health of the person, and even the frequency of the electrical current. Although the most common injuries are burns, no commercially available nonlethal force electrical device can produce a life-threatening burn.

---

a Using a 5/8-inch diameter vinyl hose that is twenty-five feet long.

Notwithstanding the admitted potential for an electrical burn, the most common accusation made against electrical devices is that the electrical shock stops the heart or causes it to fibrillate so that it twitches instead of pumping blood. Naturally, either of these conditions results in immediate death, and so blaming the use of a nonlethal electrical option for deaths that occur hours and even days later is viewed with disbelief.

Stun Guns

The term "**stun gun**"[1] refers to a number of electrical devices that deliver an electrical shock intended to disrupt muscle functions or inflict pain without causing serious injury. The distinguishing feature for these devices from other nonlethal electrical force options is that, whereas nearly all electrical force options require direct contact, these devices use one or more projectiles to engage adversaries at a distance.

By far the most well known type of stun gun is the TASER®,[2] manufactured and distributed by Axon (formerly TASER® International). Currently in use by more than 17,000 law enforcement agencies, it has been used millions of times on belligerents, as well as law enforcement and military officers during training. Although other developers have attempted to replicate the success of the TASER®, none have gained widespread acceptance. One reason is that the current models of TASERs® use compressed gas to launch the cartridge instead of an explosive, such as the primer for a lethal cartridge. These devices are classified as firearms in the United States and covered by all the associated laws and regulations.

Stun gun devices depend on an electrical shock that is of a frequency and current designed to "tetanize" the voluntary muscles, especially the arms and legs, but without causing lasting harm. This means that the muscles involuntarily contract and spasm, and the greater the muscle mass involved, the greater the effects. Pain is present; however, it is a byproduct rather than the intent. Nor is the pain in and of itself reliably adequate to attain compliance. We have observed this on many occasions when the darts impacted a belligerent too close together to affect much muscle tissue, and they continued to fight even while experiencing the pain.

I saw my first TASER® used while I was a watch commander at Lennox Station in the western part of Los Angeles. I had monitored a call that a violent person armed with a knife was threatening neighbors in an apartment building. I arrived just as my field sergeant and two deputies were entering the apartment and watched as the suspect held the knife menacingly and threatened the deputies. As he lunged toward them, the sergeant fired a TASER® that immediately dropped the suspect to the floor as the deputies grabbed his arms and handcuffed him. I thought then, and think now, that had it not been for the TASER® the suspect would have been shot.

Several years later, I was assigned to head the Los Angeles Sheriff's Department's Technology Exploration Project. This project attempted to identify, develop, and integrate the latest technologies that might have applications in law enforcement. Our focus of effort was on identifying nonlethal force options for situations that had historically required lethal force. A new TASER® manufactured by Air TASER® (later Axon) had become available and was one of my first projects. I successfully completed the instructor course, which at that time required that I get hit with a TASER® in the same manner a suspect would experience. Suffice to say that I experienced what psychologists call a "significant emotional experience."

I gritted my teeth to keep from yelling, and despite any efforts to the contrary, immediately fell to my side and rolled onto my stomach. My hands involuntarily clenched into fists and my arms pulled in toward my chest as I felt tremendous muscle tension throughout my body. I was completely immobilized and unable to recover until the electricity stopped, at which time all the effects instantly ceased. Afterwards, I felt only a deep fatigue, as if I'd been lifting great weights or running a long distance. I was exhausted. Over the years, I've been exposed to CS gas, pepper spray, stingballs, and other nonlethal options, but this was as close to being incapacitated as I had ever experienced.

Upon returning to our headquarters, I immediately briefed my Division Chief by saying, "Sir, this thing is factors more powerful and magnitudes more effective than anything we have." A four-month pilot project confirmed my experience and the LASD immediately purchased two hundred of them. Our findings revealed that the devices were effective well into the ninetieth percentile,[3] to include subjects that were mentally deranged, emotionally distraught, or under the influence of drugs and/or alcohol. Moreover, the size of the adversary was not a major factor and the devices were small enough to be worn on our gun belts. The few failures we had were nearly all attributed to flukes, such as when a dart hit the snap on an assailant's windbreaker and ricocheted away or hit the pocket of a large winter coat that contained thick gloves. TASERs® remain one of the standard items in law enforcement nonlethal arsenals throughout the country.[4]

Other Configurations and Future Plans

Nonlethal electrical force options are extremely diverse in their appearance, application, and configuration, probably because it is so easy to contrive a way to create an electrical circuit. Primarily limited only by the imagination of a developer, these options have appeared on the market as belts, rings, gloves, fences, wires, flashlights, batons, canes, pens, gloves, and just about anything else you can imagine. Some are disguised as familiar objects; others are intended to be highly conspicuous.

One variation has metal rods that extend much like an old TV antenna from a nonconductive handle. Intended for use against mobs, this item is used by touching a belligerent with both rods and allowing the electricity to flow between them. Another version uses a projectile with a capacitator that stores an electrical charge that is instantly "dumped" when it strikes a conductive surface. Some of these use small barbs to adhere to skin and clothing, whereas others have fine fibrous "wires" and small weights that complete the electrical circuit when they contact a conductor. The intent here is clearly to tetanize more muscle mass. One prototype used a thick, clear, and viscous glue-like conductive substance, nicknamed rhinoceros snot, to hold the projectile on the target long enough to deliver the electrical shock. A variation of this method was able to be launched from conventional 12 ga. shotguns and was commercially available for several years. It is no longer on the market.

Virtually all nonlethal electrical force options currently available are intended for an anti-personnel role but some, like electrical fences, can be used in an area denial role. Naturally, these need to be in place before an expected confrontation. They can be used to protect airstrips and other sensitive facilities, as well as to deter scaling walls or to reinforce gates and access points.

Other configurations include belts for combative suspects during transportation and court appearances. In this mode, the belt has no effect until remotely activated. Referred to as a stun belt, this item was developed in the early 1990s and intended for combative subjects who are already in custody but still pose a significant danger.

Still another variation is a shield with electrodes on the front that are designed to give a painful shock to anyone who comes in contact with them. The idea is to protect the user during riots and in violent custody situations, while not only increasing the ability to force belligerents away but to make the shield difficult for rioters to pull out of their hands.

Given the ease of configuration and success of existing nonlethal electrical force options, new configurations are inevitable. One patent has already been granted that replaces the wires of TASER®-like devices with an electrically charged fluid that is streamed. The major deterrence is not a lack of technology but public acceptance.

14

Biological Options

HUMANS BEGAN DOMESTICATING animals at least 20,000 years ago,[1] almost certainly beginning with the taming of wolves, the descendants of modern dogs. These dogs served early mankind in herding and as guards and hunting companions. The first domesticated horses appeared much later, around 8,000 years ago, and were probably originally used for meat rather than as beasts of burden. By about 3000 B.C.E., both horses and dogs were assisting in warfare and, to a greater or lesser extent, have participated in every war since. It should come as no surprise that they also provide nonlethal force advantages.

Besides animals that have been domesticated, animals that are inherently repulsive to humans—such as snakes, rats, hornets, cockroaches, flies, mosquitoes, lice, or skunks—can serve as biological nonlethal area denial options. They can be attracted to a specific area with the use of pheromones, and humans will then naturally move away or avoid these creatures entirely. Although it is technologically feasible to genetically alter other organisms to function as nonlethal force options, international agreements and treaties make it unlikely that these will be available in the foreseeable future.

Canines

Of all the nonlethal options available, one of safest and most effective are canines (just as often abbreviated as K-9). Used in a variety of roles, such as search and rescue, detecting contraband, tracking missing children, finding bodies, among many others, dogs' ability to apprehend violent criminals qualifies them for a place in the role of a nonlethal force option. In this capacity, they have been used to assist police since the Middle Ages. Based on successful programs in Europe, especially Belgium, the use of canines in the United States began in 1911 when New York State had sixteen dogs patrolling Long Island.[2] It is now estimated that more than 20,000 canines are in use as police service dogs today and they are an integral part of the police force in every major city in the US, as well as many countries.[3]

Serving as a nonlethal force option is a collateral duty for canines. Their primary duty is focused on finding something, especially contraband and suspects. The "battlespace"[4] for a dog is far larger and richer than for humans. They can smell things we can't, hear things we can't, and see things in low light that we can't. A properly trained dog tremendously enhances the ability of police and military personnel to recover lost children, detect drugs and explosives, and find suspects who are fleeing or hiding. It is in this last role where their value as a nonlethal force option becomes apparent.

Most US police agencies will only allow the use of canines as a nonlethal force option when encountering violent felons. A violent fleeing felon is a formidable adversary. They are in open defiance of lawful authority and are often armed with knives or guns. When caught, they are facing long prison sentences and so are desperate for escape, often resorting to armed and deadly resistance. One study revealed nearly a hundred dogs killed in a four-year period.[5] This rough average of twenty-five per year remains the same as of this writing.[6] The hazards that resulted in their deaths are the very ones their human handlers would have faced alone, and it takes no great imagination to recognize the likelihood of human deaths were it not for the sacrifice of their canine partners.

Canine Advantages

The use of canines provides several advantages over other nonlethal force options. First, their mere presence is intimidating, and many suspects will immediately surrender when they become aware that a canine will be used. This is one strong reason for providing warnings whenever practical. Typically, this intimidation factor works as a force multiplier that outperforms those of every other nonlethal option currently available.

Second, canines can strike targets in defilade. It is not unusual for suspects to try to evade the use of any force and they often take cover behind trees, cars, or other objects. A canine, however, remains focused on the suspect and will maneuver around, over, under, and sometimes through, whatever object the suspect is using to negate the effects of a force option. It is for this reason that canines are often used in conjunction with other nonlethal force options.

Third, they are the only option that can be called back after being deployed. Situations that require the imposition of any force are highly volatile and can change in an instant. When a canine handler determines that the use of a dog as a nonlethal force option is inappropriate or unsafe, they can recall the dog.

Fourth, they are the biological equivalent of **target acquisition radar**. When a suspect runs, hides, or otherwise attempts to evade, the canine quickly adapts to the new location and conditions to locate and subdue him. This is unique among all other nonlethal force options.

Fifth, canines are the only nonlethal force option that automatically adapts to increased resistance. Adversaries who attempt to fight or flee frequently suffer greater injury as the dog "regrips" to hang on or defend itself. Admittedly, it is an imprecise application of the principle but worthy of mention given the inability of any other option to adjust to meet increased resistance.

Sixth, canines are arguably the safest nonlethal force option (beyond a threat) of any other. Canines have been in use for more than a century, with countless deployments after the most dangerous criminals and under the most adverse conditions, yet only three deaths have been recorded. To be sure, dog bites are painful and nasty, but they nearly never result in death. Statistically and historically, they have proven to be the safest of all nonlethal force options currently available.

Canine Controversies

In the spring of 2001, I found myself as a newly promoted captain in charge of our Special Enforcement Bureau, which included our canine detail. Although I had worked with canines many times in the past, I had never been responsible for them. Fortunately, I had some of the best dog handlers in the United States as coaches and I took it upon myself to participate in their training and ride with them at least once a week. It was interesting, to say the least, and the coaches insisted that one of my responsibilities as the new captain was to play the role of a violent and fleeing suspect while I was wearing the heavy "bite suit."[a] It was my job to get away while fighting with the dog, and I will freely admit that I failed repeatedly. I am a fairly big guy and in reasonably good shape. I outweighed our largest dogs by double their body weight, but when they hit me at a full run, I would be tail over teakettle without the slightest hope of escape, much to the delight and amusement of the dog handlers.

This was a period in which the use of police canines had become extremely contentious and our internal policies had limited their use against only the most violent and dangerous suspects, often armed. Likewise, we only released the dog after making every reasonable attempt to obtain compliance, including warning the suspect that we would send the dog after him if he didn't surrender. We even recorded our warnings for defending ourselves against the inevitable civil suit. At the peak of this period, a Pennsylvania councilwoman gained worldwide attention when she demanded that a police dog be euthanized because the dog was racist![b] (The dog was acquitted.[7]) This allegation was only one of the many that I had to address as the new commanding officer of the unit and revealed a huge lack of understanding of the use of canines in nonlethal roles, including my own misconceptions.

One of the first things I learned is that the handler, not the dog, is the primary factor regarding when and how much force is used. The dog is a largely a product of its training. Although dogs are also individuals,

---

a This role was euphemistically referred to as the "chew toy."

b You really can't make this stuff up! Our findings revealed that dogs probably can distinguish race by smell, just like they do for individuals, but it comes without any social context. Hence, racist dogs are highly unlikely. As the years went by, we compared our bite ratios with the overall arrest statistics and discovered they were commensurate. Another strong indicator that dog racism was not a factor.

once selected, paired, and trained with a handler, the animal is more an extension of the handler than a distinct entity operating independently. As one of my handlers explained to me one time: "Sir, you are reading way too much into the dog's thinking. That dog is not agonizing over the force spectrum, the social implications, or the moral justification. To a dog, the decision is entirely on/off, black/white, yes or no."[8] Not so with the handler, who is both trained in the appropriateness of the force and can reason. Hence, the attentiveness, skill, and proximity of the dog handler have a major impact on the outcome, since they can recognize when to call off the dog. Even after the dog has latched on to a suspect, most are commanded to release in fewer than ten seconds. Notwithstanding, the use of dogs remains highly controversial for some, and has resulted in intense scrutiny and condemnation, a lack of knowledge or understanding by the critics notwithstanding.

The most common metric for measuring the "danger" of using police canines to subdue a fleeing or fighting suspect is the bite ratio. The bite ratio is simply the number of suspects bitten compared with the number of suspects apprehended. It is usually expressed as a percentage and an acceptable number varies greatly, depending on whether it is provided by those with an agenda, attorneys who make a lucrative income suing police agencies, or the courts and those responsible for the capture.

One problem is that, unlike their human counterparts, in which each use of force is intensely scrutinized and analyzed, few agencies are as heedful with force inflicted by canines. Furthermore, policies rather than circumstances have a greater effect on the number of force incidents for canines, since they impose restraints on when the use of a canine is authorized. Paradoxically, an agency that has the strictest policies and employs a canine against only the most violent and dangerous criminals is more likely to have a higher bite ratio, since these same suspects are also the most violent, have the most to lose, and tend to fight the hardest.

Another problem is that the handler is expected to participate in the apprehension of a dangerous suspect under circumstances that would justify the use of a SWAT team if he were to barricade himself. To avoid having the suspect bitten or, conversely, the canine becoming killed or injured, handlers must correctly interpret the dog's "body language," and intervene before the confrontation, often in low light and under the most adverse circumstances. At best, this is an extremely hazardous situation, and at worst, it is deadly.

A third problem is that there are three entities involved in the bite of the suspect: the suspect refusing to surrender, the canine, and the handler. Unnecessary bites can result from a poorly trained or misbehaving canine or an inattentive, lazy, or poorly trained handler.

With all that in mind, some industry standards have emerged.[9] With reasonable deployment policies,[c] a 20 percent bite ratio has been identified as likely.[10] The greater the deviation above 20 percent, the more cause for concern, and a bite ratio of less than 30 percent is an expectation.[d] The more progressive police agencies, however, look at the individual circumstances the same as any use of force by their human counterparts. Using this same "totality of the circumstances" methodology that applies to human police officers, every bite is reviewed, and any unacceptable action is cause for remedial action. This latter method is the one we (the LASD) adopted, and although we still tracked and reported bite ratios, we were guided by the specific circumstances that resulted in a bite.

Horses

Second only to canines, horses have also proven popular as biological nonlethal force options. They have served humankind nearly as long as dogs, and police agencies have used them since the 1700s. Even in the most urban of environments, horses still serve with police officers throughout the world. It takes no great leap of imagination to envision their appeal as a force option. This was made clear to me when, as a planner of the 1994 World Cup Soccer Championship games in Pasadena, California, I had our mounted enforcement unit working in a crowd-control role. These games are the most widely watched sporting event in the world

---

c Similar to "rules of engagement" for the military community.

d This methodology has been attempted to be employed for human police officers and has been strongly challenged for the same reasons. A police officer working the most dangerous, gang-infested, and crime-ridden areas can be expected to have a higher percentage of force incidents than one working a quiet residential community.

and the emotions of the spectators runs from spirited enthusiasm to rabid fanaticism. Although most in the crowds were compliant, we occasionally had incidents that erupted to near riot conditions. This was especially concerning because of a credible threat of terrorism.

One incident in particular stands out. The bus carrying the Colombian team was surrounded by a huge throng of fans, who resisted all efforts to move to safety as the bus driver attempted to maneuver the bus through the crowd to an assigned position where the team could safely disembark. After failing for nearly an hour with a full platoon of sixty police officers to move people, we called our mounted unit to assist. They quickly formed into a skirmish line with eight to ten horses, chest to chest. The squad leader then had the horses take a step forward. Immediately, the crowd moved back. The action was repeated with the same results, until some recalcitrants refused to give ground. Yet, when the command came again, the horses simply pushed them back like a biological bulldozer. Even the biggest, most determined, and defiant, whether alone or in groups, were no match for the three-quarter-ton horses, who effortlessly pushed them back. Within fifteen minutes, the bus was safely parked, order restored, and, surprisingly, people were petting the horses and having their pictures taken with the mounted unit. I don't know of any nonlethal weapon that ever enjoyed the affection displayed by the very people they were used against.

Like canines, force is a collateral duty for horses. They are primarily employed to provide a raised perspective, as well as a beast of burden to increase the mobility and logistical support available for the rider. They were especially valuable in catching auto burglars in huge parking areas and spotting anomalies, such as drug transactions, fights, fleeing suspects, and lost children in a sea of people. Moreover, they achieve success in their nonlethal force role without any allegations of animosity or viciousness.

Pheromones

Although strictly to be used for military applications, one controversial suggestion for a nonlethal biological force option is pheromones. Pheromones are chemicals that are secreted or excreted by animals, especially insects. They are used to signal things like alarm, identify food trails and swarming to a hive, mark territory, or indicate sexual availability. Because they can be synthesized, pheromones can be used to attract or repel certain types of animals and insects. When the appropriate pheromones are released in a specific area, they can be used to draw noxious animals, such as mosquitoes, bees, hornets, ants, spiders, scorpions, snakes, and the like. Understandably, humans will avoid or abandon these areas and so pheromones perform in the nonlethal role of area denial.

This idea is controversial because of the number of international treaties and conventions regarding the acceptability of chemicals and biological agents as weapons, even nonlethal ones. This option also creates ambiguity because, although chemicals are involved, they are not directly the cause of injury, and although biological agents are involved, they are those that occur naturally and even exist in the immediate area. The pheromones simply create a higher concentration. Therefore, this application does not fall neatly into either category. Because there is no practical use for law enforcement and a limited military application, there is not likely to be any viable choices for pheromones in the foreseeable future.

15

Directed Energy

NONLETHAL DIRECTED ENERGY devices are those that rely on a concentration of some type of energy that is directed over free space to affect an object or person. All directed energy is "line of sight," meaning that it travels on a straight line from a transmitter to a receiver. This means that nonlethal directed energy devices are unable to directly strike a target in defilade.[1] Examples of types of energy include light, sound, or radio frequency. They are the new kids on the block in both lethal and nonlethal force applications and so most are emerging technologies.

Light

The use of light as a weapon can be dated to at least 212 B.C.E. when the Greek mathematician Archimedes is reported to have used giant parabolic mirrors to focus sunlight on Roman ships and set them afire. Not much is known about how well it worked, but since then, light has been used as a weapon for everything from flashlights to lasers in attempting to gain some advantage over an adversary.

The most common method of using light as a nonlethal force option is by impeding the vision of a belligerent. Because of the human eye's phenomenal ability to adapt to various lighting conditions, this is highly problematic because the threshold for effectiveness is usually very close to that of being permanently disabling, especially in bright light.

Without becoming too technical,[2] let's understand that the eyes possess two predominant physiological adaptations to changes in light. The first is the constriction and dilation of the pupil. When exposed to bright light, the pupil of the human eye can constrict to less than 2mm but can dilate to as much as 9mm in low light. The larger the pupil, the greater the light energy that is able to enter the eye, where the brain then interprets it as part of an image. The pupil constricts in about half a second but can take two minutes to dilate. The second is the bleaching and regeneration of a photoreceptor protein called **rhodopsin** (also called **visual purple**). When exposed to bright light, rhodopsin immediately bleaches to a lighter, almost transparent, color, but it can take as long as thirty minutes to regain night vision. Thus, the human eye adapts very quickly to bright light but much more slowly to dim light. This concept has long been known to police officers and we frequently shine a spotlight from our radio car into the review mirror of a vehicle we are stopping at night, as well as briefly shining our flashlight into a suspect's face when we first encounter them to provide us with a tactical advantage.[a]

The Bucha Effect (Flicker Vertigo)

Besides using bright light to dazzle an adversary, pulsing the light also has a debilitating effect. This happens because the brain and the eye work together to interpret visual inputs and a flickering light can be extremely distracting, even debilitating. This phenomenon has been known since the 1940s and was studied in the 1950s in an effort to understand why some helicopter pilots reported feelings of disorientation and dizziness. The study was conducted by Dr. Bucha, who reported that when sunlight was strobed by the rotors of a helicopter at a rate that interfered with human brainwaves, people might experience epilepsy-like symptoms. The term "**Bucha Effect**" was coined to describe this condition, but it is now more commonly known as flicker vertigo.

Flicker vertigo can be duplicated by strobing a bright light between four and twenty times a second,[b] easily done with modern electronics. Critics will cite the dangers of triggering photosensitive epilepsy, in which

---

a Although mildly disconcerting, it provides a harmless advantage that gives more time to interpret the situation.

b More scientifically noted as "between 4 and 20 hertz." One hertz (Hz) is an interval of one second. Note also, that some sources expand the range from 1 to 20 Hz.

a very small proportion of the population will have an adverse reaction, to include seizures.[3] Without denying the possibility, the probability is remote, in that only about one in four thousand people are susceptible. Moreover, a plethora of other common objects—such as televisions, strobe lights, disco balls, and even sunlight reflecting off rippling waves or through a tree-lined street—can produce the same effects.

Wavelength (Color)

The wavelength of light[c] is most commonly measured in distance using nanometers (one billionth of a meter) or less often, by the frequency using terahertz (one trillionth of a second).[4] For our purposes, just know that the longer wavelengths are red and orange and the shorter ones are blue and violet. Light waves that are too long to be detected by the human eye are called infrared and light waves that are too short to be detected (on the other end of the visible spectrum) are ultraviolet.

For example, it is much harder to detect red and blue light than green or yellow light. From a debilitating perspective then, it is easier and safer to use green or yellow lights as a nonlethal force option since it is easier to overstimulate the eyes with less energy and without increasing the danger. To further increase the effects, a pulsing greenish-yellow light would seem to be most effective because pulsing light is more disorienting than a steady beam. This is now possible using colored LEDs and some developers are already experimenting with these inexpensive, low-power light sources for use as nonlethal force options.

Common Methods

Easily the best known of all the directed energy nonlethal force options is the common flashlight. Although some might question the value of a flashlight as a nonlethal force option, police officers have long understood the advantages of temporarily interfering with a suspect's sight, especially in low light conditions. Since many of our confrontations occur in low

---

c More easily understood as color.

light, this factor is especially advantageous and is routinely exploited in innocuous ways, such as the spotlight hitting the review mirror or the bright flashlight in a person's face mentioned above.

Although lasers that are eye-safe at a given distance have also been developed,[5] they are both expensive and have a minimum safe distance, which is problematic with moving suspects and inside buildings. As LEDs have become more developed and available, developers are substituting them for lasers. Not only are LEDs much cheaper, they are available in more colors and are more adaptable for installations on helmets, shields, weapons, or when worn on the body.

Sound

Like light, sound can be propagated and directed. Whereas light is part of the electromagnetic spectrum, however, sound consists of changes in pressure[d] through an elastic medium, most often the atmosphere.[e] Sound is typically measured in intensity (decibels),[6] in frequency (Hertz) and in speed (feet per second). The two effects most often used as part of nonlethal force options are intensity (loudness or volume) and frequency (pitch or tone).[7] For example, a high volume of sound will not only interfere with verbal communication but will cause anxiety and stress and even pain. A sudden burst of noise[f] can create a startle response (or **startle effect**), which is an involuntary reaction to a perceived harmful event. The response is characterized by immediate fear and increased heart rate, but more importantly for nonlethal force options, it impairs complex motor skills and interferes with the brain's information processing for as much as a minute after the exposure.

It is also well known that some noise frequencies are extremely annoying regardless of how loud they are. Probably the best-known

---

d More commonly referred to as vibrations.

e Sound is also easily transmitted through water and hard objects, such as wood and metal.

f Usually called **impulse noise**, such as the blast from a diversionary device, gunshot, or firecracker. Impulse noise is less than one second in duration and can cause permanent hearing damage without pain.

example is when someone scratches their fingernails over a blackboard. The effects are visceral and even the thought of someone doing that can cause people to cringe. Scientists who have studied this effect have noted that the frequencies are similar to a crying baby or a human scream and believe that these shrill sounds are evolutionary alarms. Measurable physiological changes in heart rate, blood pressure, and electrical conductivity of the skin are noted when people are exposed to these frequencies.[8] Because they can be artificially created and directed, it takes only a little imagination to conceive of their uses in nonlethal force for area denial or controlling mobs. Unlike tear gas, which has lingering effects, the use of noise has little or no cross-contamination or serious environmental impact and is minimally irritating to people inside buildings and vehicles.

Also, like light, the spectrum of sound is greater than just the human audible range (20–20,000 Hz), with infrasound below 20 Hz and ultrasound above 20,000 Hz. Furthermore, like the human eye with vision, human ears with hearing can adapt to an enormous range of frequency and intensity. Using audible sound as a nonlethal force option, then, is also difficult because the threshold between being effective and being dangerous is narrow.

The use of infrasound and ultrasound is, by definition, inaudible to humans but still has physiological effects, which include headaches, dizziness, fatigue, annoyance, and heart palpitations. Although initially subtle, infrasound can cause permanent effects, which include brain injuries. Ultrasound is highly directional and can be used to target small areas, especially at closer ranges. Bats, for example, use ultrasound to detect and locate flying insects and dolphins use it to create mental images of their environment. Ultrasound devices are commonly employed in medical applications, such as in sonograms for pregnancy and diagnosing maladies in soft body tissues. They are also common in motion sensors and range finders for wide-ranging applications such as sonar, car washes, and autofocus on cameras. As a nonlethal force option, ultrasound devices are already commercially available to repel animals, especially rodents and insects. Interestingly, devices are also in use to deter loitering by teenagers in a number of countries,[9] since they can hear higher frequencies than older people. These devices easily qualify as nonlethal force options performing in an area denial role.

Nonlethal Force Sound Options

One advantage of using sound as a nonlethal option is a physics factor called the **inverse square law**. Simply put, it means that as the distance is halved, the sound is doubled, and vice versa. Thus, intensity and discomfort increase tremendously the closer a subject comes to the source of the sound. Often described as a sound force field, this aspect shows great promise with proximity detectors for sensitive facilities, because the warning is intrinsic in the use of force and only the willful defiance of the adversary is necessary to increase the use of force until it becomes intolerable.

Audible sound has been used by military and law enforcement agencies as a nonlethal force option against pirates off the coast of Somalia as well as in riot control in many cities, most recently in Ferguson, Missouri, and New York City. These so-called sound cannons can range from discomforting to painful and are the sonic equivalent of tear gas, but without the environmental or residual effects.

It is extremely difficult to direct infrasound, and ultrasound has been both controversial and difficult to function within safety thresholds.[10] Currently available nonlethal force devices rely entirely on audible sound and perform primarily in the area denial role. One of the most popular has been the Long Range Acoustic Device (LRAD), which has been credited with preventing a pirate attack on a cruise ship on November 5, 2005 and has been used in law enforcement applications throughout the world, especially against mobs and rioters.

Another similar device is the Magnetic Acoustic Device (MAD). Instead of speakers, the MAD uses a diaphragm compressed between two magnets. Much more versatile than a conventional speaker, it is highly directional and provides a clear reproduction of the sound input at extremely long distances. (I have personally witnessed a demonstration at over one statute mile.) The physical structure of the MAD driver is rectangular, measuring about six by nine inches and quite flat, about three-quarters of an inch thick. Longer range devices use these drivers in combination and they can be configured to be worn on the body or mounted on a vehicle.

Radio Frequency Energy (RF)

Radio frequency energy is a type of electromagnetic radiation and is more commonly understood as radio waves. It is easily projected through free space and can be focused by the use of directional antennas. It is far more common than many people might realize and is used for radios, televisions, cell phones, microwave ovens, garage door openers, car alarms, and many other everyday items. It operates as a nonlethal force option by heating skin to the point where it becomes painful.

Since the 1980s, researchers have known that certain frequencies of radio waves create a warming effect. The invention of the microwave oven is among the first innovations resulting from this knowledge. Expanding on this research, the potential for a safe and effective nonlethal force option became possible.

Active Denial System (ADS)

The Active Denial System is a nonlethal directed energy device that can serve in both an anti-personnel and area denial role. Using a frequency of 95 GHz and an antenna to focus the beam, it can target a single individual at more than a thousand meters. It easily penetrates clothing and works by heating the skin to the depth of the nerve endings, about 1/64th of an inch (about .4mm), which causes pain but without tissue damage.[11]

In the fall of 2002, I was invited to Sandia National Laboratories in Albuquerque, New Mexico, to observe a prototype of the Active Denial System. After agreeing to be a test subject and having been briefed and completing a comprehensive health evaluation, to include a detailed interview, I stood about a kilometer away from the device housed in a CONEX box on a distant hill. I could barely make it out in the early morning sun, mostly because I knew where to look.

While waiting to begin, I faced away from the device and chatted with one of the scientists who mentioned that I was about the two-thousandth subject and that I had nothing to be afraid of. He seemed to go out of his way to put me at ease, which only increased my anxiety. And then he threw down the gauntlet when he said that just a few days previously an Army Special Forces 1st Sergeant had taken the full five-second exposure without moving, albeit even he admitted that it took a lot of grit and determination. As a Marine, I felt there wasn't anything he could have said that would make me more determined. So I decided that, "If he could do it, I could do it."

Not so patiently, I waited near a barricade that I could step behind when (if) I had had enough. To be fair, the scientist gave me a countdown so that I would not be surprised. Suddenly experiencing much more sympathy with the bull riders at a rodeo, I braced myself to uphold the honor of the United States Marine Corps and my own self-esteem. In complete silence, I soon felt a warmth on my back and clenched my teeth with as much fortitude as I could muster to withstand the five-second "ride." Less than four seconds later I was surprised to find myself in safety behind the barricade with no memory of having moved and a new admiration for the unnamed 1st Sergeant. So began my first experience with the ADS.

That was not my last exposure, and I'll candidly admit that all my future exposures ended with the same result. I have yet to withstand the full five-second ride but take comfort knowing that, with only one exception, no one else has either. I would describe the feeling as stepping into a scalding shower or someone spilling hot coffee on your back. It is not only unpleasant but will cause you to reconsider any options that don't immediately relieve the pain. What was most amazing was that the pain stopped as soon as I was no longer in the path of the beam and that there was no tissue damage, albeit I sometimes had a slight redness of my skin. After more than 13,000 test applications, only two minor injuries have occurred, with some blistering on the inside of the elbows.[12] The implications for use as a nonlethal force option were staggering.

As a nonlethal force option, the ADS is groundbreaking for several reasons. First, it is the most studied and tested nonlethal force option ever produced before being fielded. Heretofore, nearly all the data on effectiveness, not to mention safety, has been gathered from field studies.

Second, ADS is the first nonlethal force option in history to provide adequate protection against lethal force. This is because the range of the ADS exceeds the maximum effective ranges of most handheld lethal weapons, to include rifles. Third, ADS can isolate and target a single individual at extreme ranges, and so can perform as a discriminate nonlethal force option. Fourth, ADS has no adverse environmental effects. It can be used in environments that are problematic for other options, especially chemical agents. Fifth, ADS can be used against a single individual or as an "electric broom" to sweep agitators and belligerents away from one another or authorities without serious injury. Sixth, because the RF waves move at the speed of light, ADS provides instantaneous effects.[g] Seventh, ADS is extremely painful, which can be used to "teach" adversaries the consequences of defiance. Because it is invisible and silent, ADS can be enhanced with accessories such as sound and light devices to provide warnings and enable it to be used as a deterrent without requiring force!

Currently, the ADS has serious power and weight limitations that make it cumbersome to use in some applications, particularly those most likely to be experienced in the law enforcement community. There is no doubt that these will be overcome with further research and testing, and there is also little doubt that at some point in the future they will be fielded as a nonlethal force option. In fact, the major complaint so far has been that it works so well that it will be overused.[13]

The ADS is a harbinger of things to come but has aroused the ire of those who identify it as a pain ray, heat ray, and even microwave, or death ray. Whereas a case can be made for the pain and heat, the accusations of death are blatantly exaggerated. Moreover, ADS is not a microwave but rather a millimeter wave, a part of the same spectrum but orders of magnitude apart. Unlike a microwave oven, the ADS uses a "burst" of energy for only several seconds and is incapable of penetrating more than a tiny fraction of an inch below the skin.

---

g For more information on the importance of discrimination and lag time, see Chapter 8, "The Search for the Magic Bullet."

RF Vehicle Stoppers

The uses of RF energy are not limited to people and may also be used to interfere with the operation of internal combustion engines. In this manner, they would be in the anti-mobility role for nonlethal force options. Although some developers have already proven concept, a practical device remains elusive. The Joint Nonlethal Weapons Directorate has revealed a prototype capable of functioning at fifty meters with another, larger one capable of working at several hundred meters.[14] Like the ADS in the area denial and anti-personnel role, this would be a major breakthrough in nonlethal anti-mobility options. Truly a transformational capability, law enforcement would be able to choose where to stop vehicles with dangerous felons and prevent pursuits, while the military would be able to safely prevent vehicle-borne improvised explosive devices from reaching sensitive targets. Indeed, once such a device becomes available, the safety and security applications will be nearly unimaginable.

16

Conclusions

A FEW YEARS AGO, in the early morning hours on the day before Thanksgiving, in a small city on the outskirts of Los Angeles, a man carjacked a 1996 Acura Integra at knifepoint. The suspect was a thirty-five-year-old male with mental problems and a history of suicide attempts. He drove about forty miles when, just before dawn, deputies from the Los Angeles Sheriff's Department spotted him. Refusing to yield, the suspect led the deputies in a low-speed pursuit before stopping on the onramp of a major freeway. He remained defiant and refused numerous orders to exit the vehicle and surrender. Still in possession of the eight-inch knife, the suspect could be seen through the car windows angrily stabbing the seat. Fearing a physical confrontation would require deadly force, the deputies notified the Los Angeles Sheriff's Department's Special Enforcement Bureau (SWAT), who immediately responded.

Using armored vehicles to block the vehicle from moving, the SWAT team pleaded with the suspect to surrender, again without results. As the hours passed, they could see the suspect becoming more and more agitated as his breathing became heavier and he moved around inside the passenger compartment of the car while continually stabbing the seat. Trained crisis negotiators attempted to initiate a dialogue with the suspect, but he refused to even acknowledge their presence.

Finally, the SWAT deputies inserted a small canister of CS gas into the vehicle by breaking the back window. As the passenger compartment quickly filled with the tear gas, twice the suspect grabbed the gas canister and threw it out the passenger side door. However, the deputies were able to retrieve it and throw it back in the car.

When it became unbearable inside, the suspect opened the driver's door and stepped out with the knife clutched in his right hand. He was immediately struck several times with 37mm nonlethal impact munitions and fell back into the car, only to regain his footing and again exit with the knife. A police canine was deployed and grabbed the suspect by his left leg. He continued resisting and was again struck with more nonlethal impact rounds. Finally, the team leader used a TASER®, but even then, required several cycles before the suspect was able to be subdued. After being treated for his injuries, the suspect was taken into custody.

Several months later, the suspect pleaded guilty to the charges. While leaving the courthouse, a man approached one of the deputies and identified himself as the suspect's brother. He related that his brother had wanted to die and wished to personally thank the deputy and the SWAT team for not killing him.[a]

This incident serves to illustrate a number of concepts for the use of force. I chose it as a case study, not only because I was personally familiar with the circumstances and acquainted with those involved, but more importantly because it serves to highlight many aspects that would surprise those not familiar with physical confrontations and the use of nonlethal force options. Let's examine some of the main points more closely:

- Not everyone can be reasoned with, "talked down," or engaged in dialogue. The people most critical of use of force situations are nearly all inexperienced. Everything they "know" has been derived by watching TV. They are unable to comprehend a suspect's irrational actions and refusal to negotiate, even (as in this case) when a suspect repeatedly suffers the adverse consequences of defiance.

- Not all suspects are rational actors. Because of their limited knowledge, critics perceive stupid and evil acts to be logical behaviors, even when they don't understand them. On television, for example, even the most fiendish acts are scripted as part of

---

a From an interview with Scott Walker by Sid Heal (August 27, 2019) and official files of the Los Angeles Sheriff's Department.

a plot. In reality, police officers regularly encounter behaviors that are not only incomprehensible but dangerous. They react to behaviors, not motivations.

- All physical confrontations will result in pain and injuries. Critics frequently cite the injuries incurred during physical confrontations as prima facie evidence of excessive force. Here, the suspect endured one nonlethal force option after another. To illustrate the intensity of this, each strike with a 37mm impact projectile is the rough equivalent of being struck by a baseball thrown by a major league pitcher! Cuts, bruises, swelling, pain, and even dislocations and broken bones are not uncommon in physical encounters.

- Physical confrontations are not pretty. They do not look like the fight scenes on television. They more often resemble vicious dogfights rather than boxing or wrestling matches. Furthermore, there are no referees, umpires, or arbitrators. Likewise, there are no retakes, second chances, or playbacks. The force used must be decisive in nature because there is little or no time to evaluate each application to determine its effectiveness.

- The deputies attempting the arrest were genuinely concerned with the welfare of the suspect. The officers were neither excessively harsh nor indifferent to the suspect's suffering. The hours of pleading and repeated attempts to take the suspect into custody safely testify to the sincere concern that escalation could prove fatal and the primitive nature of the nonlethal force technologies utilized. However, the fact that the nonlethal options didn't work well is not as important as the fact that they worked at all! There can be little doubt that the suspect is alive today because of them.

Like it or not, we are now entering an era that history will someday recognize as the dawn of viable nonlethal force options. Admittedly, those available as of this writing are primitive and sometimes referred to as transition technologies because they are simply a stopgap as we wait for better alternatives. Using the chronology of firearms as a comparison, we are in the "blunderbuss age" of nonlethal force options. There is currently nothing available that is fully effective and incontestably safe. Those we have are short-range, inaccurate, and awkward to employ. Both failures and injuries are expected. Situations like this are replicated thousands upon thousands of times in the law enforcement and military communities. Imagine the consequences when such an encounter is multiplied by a hundred, as in a riot. Notwithstanding all of the above, the quest for safe and effective nonlethal force options must continue if for no other reason than their utilization sends an implicit message of the dignity and reverence for human life. Ultimately, history will judge us more by our yearnings and endeavors than a lack of success.[1]

Glossary of Terms

(CN) chloroacetophenone: A chemical lachrymatory agent, or just lachrymator. The term comes from the Latin *lacrima*, meaning "tear," hence the term "tear gas." CN was used almost exclusively for more than forty years, which by that time the term "tear gas" and CN had become nearly synonymous. See also *(CS), chlorobenzalmalononitrile.*

CR (dibenzoxazepine): A pale-yellow crystalline substance with a pepper-like odor. It was first synthesized in 1962 but never gained wide acceptance as a riot control agent due to a lack of supporting scientific testing.

(CS) 2-chlorobenzalmalononitrile: A chemical lachrymatory agent that also causes an itchy and watery nose and sometimes a burning sensation to the skin and tightness in the chest. Although the effects are more severe than CN, it is far safer and is preferred over other types of tear gas. See also *(CN) chloroacetophenone.*

(OC) Oleoresin Capsicum: A formulation from chili peppers, OC is used as a nonlethal option in an anti-personnel role, usually in the form of a spray. When exposed, people experience immediate and involuntary closing of the eyes, accompanied by a burning sensation and often a feeling of difficulty in breathing. Unlike CS and CN, it is also effective with mammals. See also *(CN) chloroacetophenone* and *(CN) chloroacetophenone.*

DM (diphenylaminochlorarsine): A nearly odorless, solid yellow chemical compound that causes vomiting, nausea, coughing, and weakness. Often identified as vomiting gas or adamsite after Roger Adams, the chemist who developed it. It was developed as a riot control agent but has never been popular.

Active Denial System (ADS): A nonlethal, directed energy device that uses focused radio waves in the 95 GHz spectrum to painfully stimulate nerve endings and create a sensation of heat. It was developed by Raytheon Technologies Corporation for the US military. With ranges that exceed those of most rifles, ADS is the first nonlethal weapon in history that provides an adequate defense against lethal force.

Adamsite: See *DM (diphenylaminochlorarsine)*

Adhesive Technologies: Nonlethal options that work by restricting mechanical motion. They usually consist of a chemical substance that tenaciously clings to other substances, such as skin, clothing, tires, weapons, and the like, and limits, hinders, or prevents movement.

Anti-materiel Technologies: Nonlethal options that work by affecting equipment, weapons, or supplies of an opposing force. For the present and foreseeable future, these types of nonlethal options are sought exclusively for military applications.

Anti-mobility Technologies: Nonlethal options that work by impeding or preventing the operation of a vessel, aircraft, or vehicle.

Anti-traction Technologies: Nonlethal options that work by interfering with an adversary's control over movement. Rather than restricting movement, like an adhesive technology, these frustrate it by making substances extremely slippery, resulting in uncontrollable falls and unmanageable movements. Sometime referred to as super-lubricants or super-slime. See also *Mobility Denial.*

Area denial/delay Technologies: Also known as mobility denial systems, or anti-access systems, these are nonlethal options interfere or prevent a person, vessel, or vehicle from entering or leaving an area.

Bean Bags. See *Drag-stabilized Projectiles*.

Behavior-based Rationale: A model of a force spectrum that is based upon the amount of defiance of an adversary. See also *Effects-based Rationale.*

Blackjack. See *Sap.*

Bola system: A nonlethal system that uses a lightweight cord connecting two or more weights that ensnares an adversary when thrown or launched. The name is derived from the Spanish word for ball.

Bucha Effect: Feelings of vertigo, dizziness, disorientation, and nausea resulting from strobing lights that coincide with human brainwaves. It is named for the researcher who first described it in the 1950s and is also known as flicker vertigo.

Bullwhip: A long, single-tailed whip, usually made from braided leather. As a nonlethal force option, whips work by inflicting a stinging lash.

Calmatives: Chemicals, usually in the form of a vapor, which perform as a nonlethal option by inducing drowsiness, lethargy, or sleep.

Caltrops: Devices constructed of sharpened nails, spines, or tubes, often barbed so that one projection is always up. Originally made from plants and dating from antiquity, they are typically used as a nonlethal option in an area denial role against vehicles with pneumatic tires, or mammals and people. They are also known as crowsfeet.

Capture Nets: A lightweight, flexible fabric, usually of string, rope, or wire, woven together in such a manner that it can be used to ensnare or capture an animal or human by throwing or launching it over or around them.

Cat-o'-nine-tails: A multi-tailed whip, hence the name, often with knots, bone, or other material woven into the ends. As a nonlethal force option, whips work by inflicting a stinging lash.

Chemicals: Substances and compounds formulated to cause a reaction at the molecular level. As nonlethal options, chemicals are often used in an anti-personnel role with the intent of causing a pharmaceutical interaction with a human body by injection, ingestion, inhalation, or absorption. They may also be used in anti-materiel and anti-mobility roles by interfering with internal combustion engines and other methods.

Crow's Feet. See *Caltrops*.

Defilade: An object or type of terrain feature that protects against **direct fire**.

Direct Fire: A method of engaging adversaries with extended range impact munitions by firing directly at the belligerent. This is the preferred method for employing munitions that fire a single projectile. See also *Skip-fire*.

Distraction Device. See *Flashbang*.

Diversionary Device. See *Flashbang*.

Dosage Factor: A dilemma resulting from the fact that the amount of chemical agent necessary to be effective against one person may be harmful or fatal to another, and one that is safe for all may be completely ineffective for some.

Drag-stabilized Projectiles: Also known as stun bags, bean bags, or flexible baton rounds, these are extended-range impact munitions, fired from launchers at adversaries. Drag-stabilized projectiles are those that have a "tail," much like a kite and are intended to maintain some in-flight stabilization. See also *Stun Bags*.

Durometer Scale: A scale used to measure the hardness of some material. Using a scale from 0 to 100, the higher the number, the harder the material. There is more than one scale, but the "A" scale is used for the soft rubber and plastics that typically comprise nonlethal projectiles.

Effects-based Rationale: A model of a force spectrum that is based upon the amount of injury likely to be inflicted upon an adversary. See also *Behavior-based Rationale*.

Electric Baton: A handheld baton with electrodes on one end to deliver an electrical shock in the same manner as a cattle prod.

Electricity: Energy available from an electrical charge and discharged through a conductor of some sort. As nonlethal force options, most electrical devices cause a painful shock without the trauma of impact.

Electromagnetic Energy: Waves that travel at the speed of light and contain both electrical and magnetic fields and carry energy. For use in nonlethal force options, high frequencies are focused by the use of antennae and can be used against internal combustion engines and people. Sometimes referred to as radio waves or radio frequency energy (RF).

Exposure: Vulnerability to an influencing effect. The duration of exposure is always an influencing factor when using nonlethal force options.

Fin-stabilized Projectiles: Fin-stabilized projectiles are those that use fins attached to the projectile to maintain some in-flight stabilization. See also *Drag-stabilized Projectile.*

First-order Effects: The effects caused directly by the application of force. For example, clubs cause bruising and knives cause lacerations.

Flashbang: A pyrotechnic device that, when ignited, emits a loud bang, bright flash, and pressure wave. It is also referred to as a flash/sound diversion, distraction device, or diversionary device.

Flexible Baton Rounds. See *Drag-stabilized Projectiles.*

Flicker Vertigo. See *Bucha Effect.*

Foams: A mass of small bubbles in a liquid or solid consistency. Some foams designed as nonlethal force options are closer to a liquid in consistency. Sticky foam, which works as an adhesive, resembles foam insulation. Others, such as aqueous foam, are used to hide other devices, such as caltrops. Still others are enhanced with chemical agents, such as dyes or OC, and intended to be squirted on belligerents because they cling better than liquids. Some foams are designed to be stiff and are used as projectiles.

Force Continuum: A concept used to explain the application of force through an array of options. Typically, force options are arranged by either the amount of injury likely to be suffered by the antagonist or the amount of defiance displayed by the antagonist. See also *Effects-based Rationale* and *Behavior-based Rationale.*

Forgiving Option: A nonlethal force option that has a wide margin of safety while still remaining effective.

Hypnotics. See *Calmatives.*

Impact Munition: A nonlethal projectile that performs as a nonlethal force option by striking an adversary to create pain and encourage compliance. Sometimes called an extended range impact munition.

Impulse Noise: Noise that lasts less than a second. The amount of injury likely to be incurred with noise is a combination of intensity and duration. Impulse noise can cause permanent hearing damage without pain.

Introducing Chemical Agent: There are only four ways of introducing a chemical agent into the human body. These are absorption, ingestion, inhalation, and by injection.

Inverse Square Law: A principle of physics that states that the intensity of an effect (such as experienced with sound or light) is in inverse proportion to the square of the distance from the source. More simply, as the distance is doubled the effects are halved.

Kinetic Energy: Energy involved in an object's motion. In nonlethal force applications, it's most often provided in the form of one or more projectiles.

Kinetic Systems: Any nonlethal system that relies on motion and the associated forces for effectiveness.

Knee knockers: A slang term used to describe rigid impact projectiles fired in front of an adversary and intended to ricochet off the ground or other hard surface striking them in the legs. They are usually made from wood, hard rubber, or plastic. See also *Skip-fire*.

Lasso: A long rope or cord with a noose that tightens when pulled. It is one configuration of running gear entanglement systems used to ensnare vehicle wheels or vessel propellers. See also *Loops*.

Long Range Acoustic Device (LRAD): A tightly focused sonic device that can be used to emit messages or irritating tones over long distances. It is manufactured and distributed by LRAD Corporation. See also *Magnetic Acoustic Device MAD*.

Loops: A configuration of coils in a rope, cord, or cable. It is one configuration of running gear entanglement systems used to ensnare vehicle wheels or vessel propellers. See also *Lasso*.

Magic Bullet: Any truly effective and completely safe nonlethal solution. While the criteria have been identified, the actual device remains imaginary.

Magnetic Acoustic Device (MAD): A tightly focused sonic device that can be used to emit messages or irritating tones over long distances. It's manufactured and distributed by HPV Technologies. See also *Long Range Acoustic Device LRAD*.

Malodorant: A chemical substance that performs as a nonlethal force option by releasing a pungent disgusting or repugnant odor.

Mobility Denial: Any nonlethal system that works by inhibiting or preventing movement. Whereas most are designed to restrict movement, some frustrate it by making substances very slippery, resulting in uncontrollable falls and unmanageable movements. See also *Anti-Traction Technologies*.

Net. See *Capture Nets*.

Non-stabilized Projectiles: Any projectile launched without providing ballistic advantages. Sometimes referred to as "**slick**" projectiles or just slicks.

Obscurants: Any chemical used to obfuscate or prevent vision. When used as a nonlethal force option, they often form smoke or foam. See *Aqueous Foam*.

PAVA: An organic compound and capsaicinoid that is more commonly manufactured synthetically and used in the form of a spray. In nearly every respect, it is identical to oleoresin capsicum (OC). The acronym is taken from the chemical name pelargonic acid vanillylamide; it is also known as nonivamide. See also *(OC) Oleoresin Capsicum*.

Pellets: A type of extended range impact munition that uses multiple nonlethal projectiles typically in the form of small spheres of hard rubber or plastic, which work as a nonlethal force option by striking an adversary to create pain and encourage compliance. See also *Impact Munitions*.

Pepper Spray (OC). See *(OC) Oleoresin Capsicum*.

Pheromones: Chemicals that are secreted or excreted by animals, especially insects. They are used to signal things like alarm, identify food trails and swarming to a hive, mark territory, or indicate sexual availability. They perform in a nonlethal area denial role by attracting noxious, poisonous, or dangerous animals.

Plume Cloud: A relatively elongated cloud emitted from a single source, as smoke from a fire. The name comes from the shape inspired by the plume of a fire. In a nonlethal context, it is often used to describe the deployment of tear gas or obscurant smoke.

Projectile: Any object thrown, flung, or launched through the air.

Pseudo-symmetry: A misleading perspective that occurs when opposing views are presented as equally valid, but one lacks scientific authority. It occurs in the media when it appears to viewers and readers that subject matter experts are equally divided on an issue but, in actuality, the vast majority holds a differing opinion.

Radio Frequency Energy (RF): See *Electromagnetic Energy*.

Rhodopsin: A light-sensitive protein found in the eye. When exposed to light, it "photobleaches" and reduces the amount of light energy. Conversely, in low-light conditions, it gathers light energy to enable night vision. Although it can reduce the light energy in less than a second, it can take as much as thirty minutes to regenerate to allow night vision. It is also called visual purple.

Rifled: The internal spiral grooves that cause projectiles to spin during flight. Spinning greatly improves ballistic stabilization and accuracy. It is also referred to as a rifled barrel or rifled bore.

Ring Airfoil: A projectile that resembles an aircraft wing folded into a donut-shape and provides in-flight lift and stabilization.

Riot Control Agents: Any nonlethal chemical agent used for riot control. The most commonly used are pepper spray (OC) and tear gas, usually CS. See also *Tear Gas*.

Risk Exposure: A vulnerability to an adverse event resulting from being susceptible to harm.

Risk Probability: An estimation of the likelihood of some misfortune or setback. It is expressed as either a percentage or a ratio.

Risk: The susceptibility of experiencing a setback or harm during an unfolding event or in contemplating a course of action. Risk has two components, probability and exposure. See *Risk Probability* and *Risk Exposure*.

Rubber Bullets: A slang term used to describe nonlethal projectiles.

Running Gear Entanglement Systems (RGES): Any device employing ropes, cords, or cables and designed to foul propellers or entangle the wheels and powertrain of a vehicle. Some also use adhesives in the same manner. They are used as nonlethal force option in an anti-mobility role.

Sap: A small, handheld impact weapon made of leather and filled with lead shot. They are used as nonlethal force in an anti-personnel role. They are also called blackjacks, slappers, or **slapjacks**.

Second-order Effect: The effects that occur from the use of force indirectly, such as from falling. Although the root cause is the application of the force, the proximate cause is the circumstances and environment in which the force was applied.

Sedatives. See *Calmatives*.

Shaping Operation: Any series of actions taken in anticipation of an engagement or tactical operation designed to promote accomplishment of strategic objectives.

Shock Belt: A belt fastened to a subject's waist, leg, or arm that emits a strong electrical shock when a remote control is activated. Sometimes (erroneously) referred to as a stun belt.

Sjambok: A relatively short but heavy leather or plastic whip. It was used as a nonlethal force option during the Apartheid era by the South African Police Service and became a symbol of oppression.

Skip-fire: A method of engaging adversaries with extended range impact munitions by firing at an object, usually the ground, and "skipping" the projectiles into the target. This method has proven safer and more effective than direct firing when a munition relies on multiple projectiles. See also *Direct Fire* and *Knee Knockers*.

Slapjack. See *Sap*.

Slick Projectiles. See *Non-stabilized Projectiles*.

Somnolents. See *Calmatives*.

Soporifics. See *Calmatives*.

Spike Strips: A nonlethal device using metal spikes, often hollow, to puncture the pneumatic tires of a vehicle driven over them. Some variations allow for the hollow spikes to detach and embed themselves in the tires to hasten the deflation. They are used as nonlethal force options in an anti-mobility or area-denial role.

Spin-stabilized Projectiles: Nonlethal projectiles that spin in flight to establish and maintain ballistic stability. Some use fins while others are fired from a rifled barrel. See also *Drag-stabilized Projectiles*, and *Fin-stabilized Projectiles*)

Sponge Grenades: Soft-nosed projectiles fired with a launcher, most often a 37 or 40mm. They are intended for direct fire and are highly accurate, with a longer range in comparison to other nonlethal weapons, and they are one of the most consistent in energy transfer. The name is taken from the sponge-like material at the front of the projectile. They are also known as sponge rounds.

Standards by Consensus: Refers to those standards accepted by the majority of the practitioners. Although there is no legal requirement to follow these types of standards, they become a de facto standard of comparison and so are highly influential. See also *Standards by Mandate*.

Standards by Mandate: Refers to those standards required by rules of engagement, law, or agency regulation. They are the default method of imposing requirements. See also *Standards by Consensus*.

Standoff: The distance between adversaries, often established by the weapons at hand.

Startle Effect: The involuntary physiological effect that occurs when a person is alarmingly stimulated. It is used to sight invisible weapons, such as those relying on sound, invisible light, or radio frequency energy.

Sticky Foam: An expansive chemical that works as a human adhesive. It can also be used against some vehicles and provides nonlethal advantages in anti-personnel or area denial roles.

Stingball: A pyrotechnic device that looks like a black rubber softball, but explodes and flings small, hard rubber pellets that sting when they hit, hence the name. Some also contain powdered chemical agents with the pellets.

Stun Bags: Also known as bean bags or flexible baton rounds, these are extended-range impact munitions, fired from launchers at adversaries. The most common configuration is a small sack, filled with lead pellets, sand, or other similar material, intended to create pain upon impact but without serious injury. They are predominantly used in a nonlethal, anti-personnel role. See also *Drag-stabilized Projectiles*.

Stun Gun. See *TASER®*.

Super Slime. See *Anti-traction Technologies*.

Super-lubricant. See *Anti-traction Technologies*.

Swett Curve: A method of plotting statistical data in which successive categories are added to the previous ones. A normal distribution is displayed as a "lazy S." It takes its name from Charles Swett, who demonstrated its usefulness in identifying risks of nonlethal force options.

Target Acquisition Radar: Refers to radar that automatically detects, locates, and acquires targets. Although the actual devices are limited to lethal applications, the concept is sometimes used in nonlethal descriptions and concepts.

TASER®: A nonlethal device that fires barbs attached to wires to deliver a shock. TASER® is an acronym, and so capitalized. Unlike most other nonlethal electrical devices, TASERs® work by the involuntary constriction of muscle tissue (tetanization) rather than pain. Although pain occurs, it is a side effect rather than the intent. Sometimes (erroneously) called a stun gun.

Tear Gas: Any of a number of chemical agents that are used as nonlethal force options and work by causing the eyes to "tear." Although often used as riot control agents, they are also available as sprays and aerosols, for individual combatants and barricaded suspects. See also *Riot Control Agents*.

Technicals: A militia in Somalia that armed themselves with heavy weapons mounted in the beds of pickup trucks.

Tetanization: The involuntary spasm of muscles resulting from a chemical or electrical stimulus. See also *TASER®*.

Third-order Effects: The effects of an application of force that exacerbates an existing health condition. These effects are neither directly related to the force nor dependent upon the circumstances or environmental factors, but rather inherent in the unique health conditions of the subject.

Visual Purple. See *Rhodopsin.*

Vomiting Gas. See *DM (diphenylaminochlorarsine)*.

Notes

All URLs were accessed as of April 21, 2020.

1: Nonlethal Confusion and Controversies

1 Policy for Nonlethal Weapons, US Department of Defense Directive number 3003.3, July 9, 1996. The original policy has been superseded by newer revisions but other than removing the hyphen, the term and definition have remained intact. Available at the Homeland Security Digital Library at https://www.hsdl.org/?abstract&did=464197.

2 In the same manner, "nonlethal" weapons can be used to kill or seriously injure an adversary. Examples include head strikes with batons and saps, targeting the throat or face with impact munitions, and so forth.

3 He could have just as accurately replied, "It depends on who is using it." This strikes to the heart of a related issue, that of using lethal force for protection against a suspect who is in control of a highly effective nonlethal force option, such as a TASER®. Given appropriate circumstances, a knowledgeable officer could reasonably infer that a suspect would be able to use such a device to gain control of his service weapon and then kill him with it. It all revolves around the intent.

4 One civil rights group objects to any kind of advance warning by demonstrating a nonlethal device to "assist" a belligerent in making an informed decision because it intimidates them. My typical reply has been that, at worst, no harm is done and, at best, no force is necessary. From a street cop's perspective, I don't see any downside to this common practice.

5 Operation United Shield was conducted in the spring of 1995 by US Marines and Naval personnel and assisted by the Pakistan and Italian navies. Spearheaded by USMC Lt. Gen. Anthony Zinni, it was the first military operation in history to incorporate nonlethal force at all echelons of the conflict and successfully evacuated UN Peacekeeping Forces from Somalia with a minimum of casualties. This operation is often cited as the forerunner of the interest in and efforts of modern nonlethal programs.

6 See the Glossary of Terms for some of the working definitions from various organizations and countries.

7 Policy for Nonlethal Weapons, US Department of Defense Directive number 3003.3, July 9, 1996.

8 For a more in-depth view of the issue see: Ed Hughes (ed.), "Minimal Force Options and Less-Lethal Technologies." *Report of the Third International Law Enforcement Forum*, Home Office, United Kingdom; Applied Research Laboratory,

Pennsylvania State University; Institute for Nonlethal Defense Technologies (3–5 February 2004), 182.

9 In November 2016, water was sprayed by police in frigid temperatures near Bismarck, North Dakota to disperse some 400 protestors during a demonstration against the Dakota Access Pipeline.

10 This is expected to change in the near term as demands for more information from the American public increase. In January of 2019, the FBI initiated a database to track deadly force and serious bodily injury incidents involving police. Each year, there are an estimated sixty million encounters between police and citizens with an estimated thousand deaths. This demand is largely seen as a welcome development by local law enforcement.

11 This includes unintended effects, such as scarring, carcinogenic, mutagenic, or any other residual effects.

2: A Brief History

1 Adlai Stevenson, in a speech at Richmond, Virginia, September 20, 1952.

2 The Biblical account is recorded in the Book of Joshua, Chapter 20. For a more thorough examination of the account specifically relating to the employment of nonlethal weapons, however, see Col. John B. Alexander, *Future War: Non-Lethal Weapons in the Twenty-First Century* (New York: St. Martin's Press, 1999), 95.

3 In point of fact, this "salting of the earth" was a method dating back at least a thousand years further. There are a number of accounts occurring in the Fertile Crescent, where Assyrians and Israelites salted cities for the same purpose.

4 Alexander, *Future War*, 77.

5 Adrienne Mayor, *Greek Fire, Poison Arrows & Scorpion Bombs* (New York: Overlook Press, 2003), 225.

6 Of note, however, is that uses of electrical devices have been reported in South America, notably Argentina, as early as 1932. They quickly became used as instruments of torture, however, and whereas some might claim that they still meet the modern definition of a nonlethal weapon, I would submit that they fail the test in spirit. Consequently, I have intentionally ignored them. For those interested in a more comprehensive perspective of the history of nonlethal weapons, I would highly recommend reading "The Early History of 'Non-Lethal' Weapons," Occasional Paper No. 1, and "The Development of 'Non-Lethal' Weapons During the 1990s," Occasional Paper No. 2, both by Neil Davison, Bradford Non-Lethal Weapons Research Project (BNLWRP), Department of Peace Studies, University of Bradford, Bradford, UK, December 2006 and March 2007, respectively.

7 Eugene Nielsen, "The Selection and Tactical Employment of Less-Lethal Chemical Munitions: A Basic Overview," *Police and Security News* 13, no. 6 (November/December 1997): 41–46.

8 Anna Feigenbaum, "100 Years of Tear Gas," *The Atlantic*, August 16, 2014, https://www.theatlantic.com/international/archive/2014/08/100-years-of-tear-gas/378632/.

9 One possible exception, largely attributed to folklore, is the use of rock salt fired from a shotgun used by rural civilians to dissuade trespassers. Although there is almost certainly some underlying truth, I did not include it since I could find no

instances that it was ever used by legitimate authority, accounts were anecdotal, the munitions were never available commercially, and the incidents were historic. The reports, however, were all in recent times.

10 Brian Rappert, *Non-Lethal Weapons as Legitimizing Forces? Technology, Politics and the Management of Conflict* (London: Frank Cass Publishers, 2003), 38. See also, Davison, "The Early History of 'Non-Lethal' Weapons," 10.

11 Between 1970 and 1999, more than 126,000 baton rounds were fired in Northern Ireland resulting in at least seventeen deaths and more than 600 injuries. As training, experience, and technology continued to improve, the last death was recorded in 1989. (Numerous personal interviews with Colin Burrows, Chief Superintendent (ret.), Northern Ireland Police Service, 2007–2008). See also: N. Lewer, (ed.), *The Future of Non-Lethal Weapons: Technologies, Operations, Ethics and Law* (New York: Routledge, 2013), 99–111.

12 Davison, "The Early History of 'Non-Lethal' Weapons," 10.

13 Cross-contamination refers to the likelihood of affecting persons other than the intended target. In a spray form, both CN and CS are notorious cross-contaminators and many police officers avoided using them against combatants, who were often less affected than the police.

14 Even the most conservative estimates place pepper spray as a nonlethal option in more than 95 percent of American law enforcement agencies. In a wide variety of dispensers, from key chains and flashlights to belt clips and lipstick cases, pepper spray is also legal and available to citizens throughout the United States.

15 The Bureau of Justice Statistics reports that 94 percent of local police agencies in the United States authorize the use of pepper spray and 87 percent authorize the use of batons. See Brian A. Reaves, Ph.D. "Local Police Departments, 2013: Equipment and Technology," Bureau of Justice Statistics, US Department of Justice, Office of Justice Programs, July 2015, NCJ 248767 https://www.bjs.gov/content/pub/pdf/lpd13et.pdf.

16 PAVA is also known as nonivamide. Like OC, PAVA is commonly dispersed in either a powder or liquid form.

17 John Barry and Tom Morganthau, "Soon, 'Phasers on Stun,'" *Newsweek*, February 6, 1994, http://www.newsweek.com/id/113112.

18 Davison, "The Development of 'Non-Lethal' Weapons During the 1990s," 13.

19 These are made from about forty grams of #9 lead birdshot sewn into a Cordura® or ballistic nylon pouch and launched from a 12-gauge shotgun with a cylinder choke. Whereas modern derivations may vary to some degree, by and large they conform to this general design.

20 The term TASER® is an acronym representing the name given by the inventor, Jack Cover. Cover, a NASA researcher, had a working device as early as 1974 and called it "Thomas A. Swift's Electronic Rifle," after the popular hero of juvenile adventure novels.

21 On March 3, 1991, a vehicle driven by Rodney King, after evading the California Highway Patrol on a freeway for several miles before driving onto surface streets, finally stopped. During King's arrest, he resisted being handcuffed and was twice TASER®-ed without effect. Eventually, he was struck numerous times with batons until subdued. The vehicle stop was filmed by a bystander and the video attracted international attention. Four of the

officers were tried and acquitted, resulting in massive rioting. Eventually, the officers were charged under federal laws and two of them were found guilty. An independent commission (the Christopher Commission) was created to examine training practices, internal discipline, and citizen complaints. The incident is largely credited with creating the public's intense interest in police use of force and set the stage for an increased demand for more advanced nonlethal force options.

22 TASERS® are now used by more than 17,000 law enforcement agencies in more than a hundred countries.

3: So! Why Bother?

1 The type of force can be a particularly sensitive issue when a particular option is accompanied by cultural, racial, or historical issues. Examples include the misuse of dogs by American law enforcement during the civil rights demonstrations in the 1960s or the use of sjamboks (a heavy leather whip, often made from the hide of a rhinoceros or hippopotamus) by the South Africans during Apartheid, both of which became symbolic for oppression.

2 Paraphrased from Abraham H. Maslow, *The Psychology of Science* (New York: Harbor & Row, 1966), 15. Maslow called this "the law of the instrument" or "the law of the hammer."

3 Those aligned with Mohammed Farah Aideed.

4 This is the infamous "Black Hawk Down" incident, and is also known as the First Battle of Mogadishu, to distinguish it from the eight more battles that followed.

5 United Nations Security Council Resolution 954. This also marked the first time the United Nations had left a country without attaining its objectives.

6 Most notably, Pakistan and Italy, who each provided two ships and assisted in the evacuation of the UNISOM troops, primarily from Bangladesh, Pakistan and Italy.

7 Policy for Nonlethal Weapons, US Department of Defense Directive number 3003.3, July 9, 1996, available at the Homeland Security Digital Library at https://www.hsdl.org/?abstract&did=464197.

8 Because situations that will benefit from nonlethal options tend to escalate, coupled with the relatively primitive nonlethal options available, an ability to intervene early is paramount. Consequently, both the availability of the nonlethal weapons *and the authority to use them* must be pushed to the lowest possible levels of the organization.

9 Napoleon I, *Maxims* (1804–05), "All becomes easy when we follow the current of opinion; it is the ruler of the world."

10 During Operation United Shield in Somalia, we euphemistically called this advantage an "irrefutable, cross-cultural, language-independent signal of sanction." Although the term was eventually polished by all of the members of the nonlethal Mobile Training Team (MTT), the real author was (then) Lt. Rob Ireland, our laser expert from the USAF. He was smarter than all the rest of us put together and was a key contributor to the success of our efforts.

4: Using and Justifying Force

1 The US military typically uses the term "escalation of force."

2 One Marine Corps officer explained to me in a passageway of the USS Belleau Wood during Operation United Shield, spring of 1995, that the force spectrum for the USMC was "M-16 to F-16. There is no room for nonlethal options."

3 USMC Col. Kirk Hymes, the Director of the US Department of Defense's Joint Non-Lethal Weapons Program, calls this dilemma "shout or shoot."

4 In other tactical settings, the term "force" can also be used as a noun to describe a body of troops, such as a military force, police force, or labor force.

5 By no means is the complexity limited to just these noted aggravations. Others include, allegations from detractors that because a nonlethal option does not incur the same adverse consequences as lethal force it tends to be used in place of milder alternatives. Human nature being what it is, this is probably true. The US courts have ruled, however, that there is no necessity to use the minimal force, only that the force used was reasonable under the circumstances. The landmark case on this is from the US Supreme Court: Graham v. Connor, 490 U.S. 386 (1989). Another common allegation is that nonlethal weapons can easily be used for torture. TASERs® are among the most commonly cited in this regard. If a user is predisposed to torture someone, this might be true. The logic is shallow, however, in that how any force option is applied will always be somewhat controversial. Blaming behavior on an inanimate object, such as torture on a TASER®, is like blaming obesity on food. Besides, as noted expert Dr. John Alexander observes: "Why would you pay $800 for a TASER® when a $10 pair of pliers will work just as well?"

6 One question on threats that often arises concerns warning shots. Even though warning shots require the use of a "deadly" weapon, they are, by definition, nonlethal. Remember, it is not the capability but the *intent* that is the principle division for separating lethal and nonlethal force. Thus, lethal weapons can be used in the delivery of nonlethal force (as in a "shot across the bow") and nonlethal weapons can be used to deliver lethal force (as in head strikes with a baton). Accordingly, a credible threat of lethal force is a nonlethal option.

7 Although this is a good example of a force continuum, it is by no means definitive. The US military uses one with five distinctive steps: cooperative controls, contact controls, compliance techniques, defensive techniques and deadly force.

8 US Supreme Court: Graham v. Connor, 490 U.S. 386 (1989), https://supreme.justia.com/cases/federal/us/490/386/.

5: Injuries from Nonlethal Force Options

1 It is worth mentioning that there are virtually no circumstances that will preclude injuries from nonlethal force.

2 One of the earliest to evaluate injuries to both suspects and officers with nonlethal weapons was (then) Capt. Greg Meyer of the Los Angeles Police Department who studied a number of force incidents in which all manner of nonlethal weapons were used. As might be imagined, those nonlethal options that required the closest proximity to the suspect resulted in notably higher injuries to the user. Incidents in which a baton was used, for one example, resulted in 15 percent

of the users (officers) sustaining moderate to major injuries. See Greg Meyer, "Nonlethal Weapons versus Conventional Police Tactics: The Los Angeles Police Department Experience," master's thesis, California State University, Los Angeles, March 1991, Table 4.7.

3 One study showed that, overall, the chances of a user being injured at the second iteration ("dose") was about three percent but rose to 11 percent at the third iteration, a nearly fourfold increase! Furthermore, injuries to both officers and suspects rose commensurately with the length of the confrontations. See Charlie Mesloh, Mark Henych, and Ross Wolf, "Less Lethal Weapon Effectiveness, Use of Force, and Suspect & Officer Injuries: A Five-Year Analysis," A Report to the National Institute of Justice, Florida Gulf Coast University, Weapons & Equipment Research Institute, September 2008, 68 and 90, https://www.ncjrs.gov/pdffiles1/nij/grants/224081.pdf.

4 It has been concluded that impact munitions can cause forty-seven different injuries; but four predominate. These are fractures, concussions, lacerations, and organ damage. Even these represent a very low percentage, however, with less than two percent of the total associated with these types of munitions. See Dr. John M. Kenny, *et al.*, "Non-Lethal Weapon Injury Characterization," Human Effects Advisory Panel, Institute for Non-Lethal Defense Technologies, Penn State, State College, PA (August 30, 2004), 12.

5 *Ibid.*, 17–18.

6 This solution was first proposed by the Royal Ulster Constabulary (now the Police Service of Northern Ireland), which had the experience of firing about 126,000 of these projectiles resulting in seventeen deaths from 1970 through 1989, but none afterwards. In addition to changing the aiming point, the solution includes consistent performance from the munitions and an accurate, ergonomic launcher. See Colin Burrows, "Operationalizing Non-Lethality: A Northern Ireland Perspective," in *The Future of Non-Lethal Weapons: Technologies, Operations, Ethics and Law*, 2nd ed. (London and New York: Routledge, Taylor & Francis Group, 2013) 101–104, and numerous personal interviews and exchanges of email with Burrows, Chief Superintendent (Ret.), Police Service of Northern Ireland (formerly Royal Ulster Constabulary).

7 Dr. John M. Kenny, *et al.*, "The Attribute-Based Evaluation (ABE) of Less-Than-Lethal, Extended-Range, Impact Munitions," Pennsylvania State University, Applied Research Laboratory and Los Angeles Sheriff's Department, State College, PA (February 15, 2001), 2 and 46.

8 It should be noted that these agents are not a gas in the conventional understanding. CN and CS are solid compounds that are micro-pulverized to allow dispersal with smoke or wind and initially appear as a translucent gray cloud, hence their appearance as a gas. Because they are heavier than air, they eventually fall to the ground. If sufficient quantities remain in an area, they can be revived if picked up by the wind or from people walking through it. Although not as commonly used in riot control, OC is a finely ground residue from pepper plants and relatively heavy when compared with either CN or CS, and so falls to the ground more quickly. It is highly effective against individuals; however, its weight makes it more difficult to use as an area contaminant. It is,

however, sometimes combined with other agents, especially CS, to exploit the advantages of both.

9 There are some handheld dispensers with greater ranges, as much as thirty-five feet, but these are large cylinders with hoses and are used in support of other nonlethal options, especially during riot situations, rather than as individual weapons.

10 Both personal experience and scientific studies have shown that many intoxicated suspects will react violently when sprayed with OC. See Charlie Mesloh, Mark Henych, and Ross Wolf, "Less Lethal Weapon Effectiveness," *supra* note 3, 37.

11 For brevity and clarity, this chapter only describes those measures necessary for users to make effective deployment decisions. Of necessity, some options have been omitted. Although calmatives and soporifics are still a novelty, their use in the Dubrovka Theater raid in Moscow is certainly a harbinger of things to come. Consequently, it needs to be said that the tragic deaths of hostages from the soporific agent employed was not entirely the fault of the chemical agent itself and implementing some protective measures can avert similar deaths in the future. After being affected by the gas, some of the victims died of asphyxiation whereas others choked on their own vomit. Many languished for some time without treatment. Medical treatment was delayed, and when it became available, medical personnel were unaware that a soporific had been used or that a counteragent was available. Had it been administered, it might have saved lives. Two life-saving lessons might be gleaned from this tragedy. First, it is far more efficient to bring the medical personnel to the victims rather than try to move victims to medical facilities. This was a critical lesson learned during the Oklahoma City bombing in 1995 and has become common practice in emergency responses throughout the United States. Secondly, a fully informed medical staff is far more capable of gathering the necessary equipment and supplies and administering counteragents than those required to diagnose an unconscious patient.

12 In point of fact, there are always first-order injuries, but they are so minor as to be trivial. Although I'll be the first to admit that I don't relish having sharp-barbed probes embedded in my flesh, or the redness and sometimes small blistering from electricity passing through my skin, these injuries are actually quite insignificant and nearly all heal completely with no scarring within two weeks.

13 In point of fact, there are too many to list, but a brief examination of the Internet will provide more than enough corroboration. One of the more recent studies summarizes the findings as follows: "Although exposure to CED [conducted energy devices] is not risk free, there is no conclusive medical evidence within the state of current research that indicates a high risk of serious injury or death from the direct effects of CED exposure." National Institute of Justice, *Study of Deaths Following Electro Muscular Disruption: Interim Report*, Washington DC (June 2008), 3, http://www.ncjrs.gov/pdffiles1/nij/222981.pdf.

14 Although police dogs are trained specifically to grab and hold on to extremities, the injuries suffered by untrained dogs also tend to be localized to arms and legs. Emergency rooms report that approximately two-thirds of dog-related injuries are punctures or lacerations. See J. Gilchrist, MD, Division of Unintentional Injury Prevention; K. Gotsch, MPH; JL. Annest, PhD; G. Ryan, PhD. Office

of Statistics and Programming, National Center for Injury Prevention and Control, "Nonfatal Dog Bite—Related Injuries Treated in Hospital Emergency Departments—United States, 2001," Centers for Disease Control and Prevention, *MMWR Weekly* 52, no. 26 (July 4, 2003): 605–610, http://www.cdc.gov/mmwr/preview/mmwrhtml/mm5226a1.htm.

15 Although severe injuries are not acceptable, virtually no other nonlethal option can make that claim. Thus, the use of canines is one of the safest nonlethal options available. See Charlie Mesloh, Mark Henych, and Ross Wolf, "Less Lethal Weapon Effectiveness," *supra* note 3: 33–34. See also: Robinette v. Barnes, 854 F. 2d 909 (Sixth Cir. 1988), https://law.resource.org/pub/us/case/reporter/F2/854/854.F2d.909.86-6135.html.

6: Assessing and Managing Risk

1 Although the verbiage may sound different, both law enforcement and military rules of engagement justifying lethal force are notably similar. In somewhat over-simplified terms, lethal force is authorized for law enforcement officers to save their life or the life of another. Military verbiage uses terms like "hostile act," "hostile intent," or "hostile force," but self-protection is inherently and universally understood as acceptable. Both disciplines evaluate the necessity of deadly force on two fundamental criteria: a reasonable belief that it was necessary, and an adversary with the present ability to carry it out.

2 Although pain is present with both the use of a conducted electrical device (e.g. TASER®) and tear gas, it is more of a byproduct rather than the intended consequence. Conducted electrical devices employ muscle tetanization, whereas tear gases (including pepper spray) blur vision; cause stinging and irritation of the eyes, nose, mouth, and skin; and induce coughing.

3 Charles Swett was a policy assistant in the Office of the Assistant Secretary of Defense for Special Operations and Low-Intensity Conflict. He was an early proponent of nonlethal options and has lectured and written extensively on the subject. He originally called it "S-curves" after the typical form of the cumulative frequency polygon, but it was quickly dubbed the "Swett Curve" by those who knew the origination. It has been in use since around 1996 (interviews via email with Charles Swett, January 22, 2009).

4 I know I'm faltering in my attempts to keep the information simple, but it is no less important. A cumulative frequency distribution is simply a graph that displays the running total of the scores. In this case, it is showing the frailty and vigor of the human population in comparison with force. Think of it as the typical bell-shaped curve you're more familiar with, but with the successive values added to the preceding ones instead of spread out along the x-axis.

5 The figures are only for illustrative purposes. The proximity of the three curves to one another is directly related to the type of nonlethal force being examined. For a device that uses light stimulation, such as a laser dazzler or light emitting diode (LED), the effective threshold lies very close to the permanently disabling threshold, indicating the small safety margin when attempting over-stimulation without harm, whereas the lethal threshold remains quite far apart. If we were to examine pepper spray, the thresholds for permanently disabling and lethal would be closer together but quite far away from the effective threshold, indicating

the much wider margin of safety. Nonlethal options that have wide margins of safety are called "forgiving." The Swett Curve is one of the most useful tools for comparing disparate nonlethal options that may take a wide variety of forms such as liquids, sprays, projectiles, chemicals, electricity, electromagnetic energy, and so forth.

6 Nonlethal options that attempt to overstimulate sight and hearing are typically the most dangerous. This is because the human eyes and ears are so adaptable to changing conditions that it is nearly impossible for a force option to be effective without approaching or even exceeding the thresholds for permanent injury.

7 The likelihood of hitting a man-sized target at a given range was an early standard of measurement with nonlethal projectiles. In some places, it still is. In comparison, the standard for nearly every lethal projectile is "point of aim" equals "point of impact." This creates a blurry standard because unintentionally hitting a person in the throat or eye when a projectile was not accurate has allowed the developer or manufacturer to claim it "performed to standards." For a more comprehensive examination of this subject see Dr. John M. Kenny, *et al.*, *The Attribute-Based Evaluation (ABE) of Less-Than-Lethal, Extended-Range, Impact Munitions*, Pennsylvania State University, Applied Research Laboratory and Los Angeles Sheriff's Department, State College, PA (February 15, 2001).

8 It should not go unmentioned that risk competes with tactical objectives, and decreasing probability or eliminating exposure may not be practicable.

7: Voodoo Science and the Media

1 *A New Beginning: Policing in Northern Ireland*, The Report of the Independent Commission on Policing for Northern Ireland, September 1999, 54, http://cain.ulst.ac.uk/issues/police/patten/patten99.pdf.

2 Press Release. ACLU Urges CA Appeals Court to Declare Use of Pepper Spray Dangerous and Cruel, August 12, 1999, http://www.aclu.org/police/abuse/14551prs19990812.html.

3 Pepper Spray Update, ACLU of Southern California June 1995, https://kulturlerarasinda.files.wordpress.com/2013/07/pepper-spray-update-more-fatalities-more-questions.pdf.

4 In all fairness, scientific measurements for effectiveness of nonlethal force options are not well established. Indeed, even the definition of effectiveness is not universally accepted. Other studies, using different metrics, are substantially lower. What is irrefutable, however, is that TASERS® are one of the best options when compared with other force alternatives when safety, injuries, and intended results are compared.

5 Although it has also been called "junk science," the term "voodoo science" is taken from the excellent book *Voodoo Science: The Road from Foolishness to Fraud* by Robert Park (New York: Oxford University Press, 2000).

6 The group is defunct; however, remnants of their website remained for years afterwards. Sunshine Project http://www.sunshine-project.org/ accessed on January 24, 2009.

7 As of this writing, the latest figures come from Reuters, reporting 1,081. Tim Reid, Peter Eisler, and Grant Smith, "As Death Toll Keeps Rising, US Communities Start Rethinking TASER® Use," Reuters, February 4, 2019, https://

www.reuters.com/article/us-usa-taser-deaths-insight/as-death-toll-keeps-rising-us-communities-start-rethinking-taser-use-idUSKCN1PT0YT. This figure is a running tally offered by various civil rights and news organizations.

8 The figure is taken from a continually updated algorithm based on a study in "Field Statistics Overview" by James E. Brewer, *et al.* In TASER® *Conducted Electrical Weapons: Physiology, Pathology, and Law* edited by Mark W. Kroll and Jeffrey D. Ho (New York: Springer, 2009), 283–300.

9 When this fact is pointed out to some, a common response is, "These people are healthy and not under the influence of drugs or alcohol." How do you respond to that? Note: This figure is a conservative estimate based on reports and derived from an algorithm. It is considered to be conservative because many exposures in this group are never reported.

10 As the years have passed, I frequently update my statistics with the latest figures and reports of deaths, especially before speaking to audiences or the press. It is an exceptionally rare occurrence when the risk ratio has deviated by more than 1/100th of a percent!

11 Likewise, the distortion of perspective has occurred in the UK, where it is routinely reported that at least seventeen people have died from being struck with baton rounds in Northern Ireland. What is nearly never reported is that advanced technologies and better training have drastically reduced the dangers and that the last death occurred in 1989! Numerous personal interviews and exchanges of email with Colin Burrows, Chief Superintendent (Ret.), Police Service of Northern Ireland (formerly Royal Ulster Constabulary) See also: N. Lewer, (ed.), *The Future of Non-Lethal Weapons: Technologies, Operations, Ethics and Law* (New York: Routledge, 2013), 99–111.

12 This is an issue where the law enforcement and military communities part company, in that inaccuracies and mischaracterizations reported in the media rarely go unchallenged by law enforcement but are seldom even commented upon by the either the military or Department of Defense. To be sure, there are times to "refuse battle" but when critics distort facts and monopolize public opinion through the media it also surrenders the field.

13 For an excellent examination of this issue see, Dr. John Alexander, "The Need for Advocacy: A Proposal," *Proceedings of the 5th Symposium on Non-Lethal Weapons*, Ettlingen, Germany (11–13 May 2009), 0–1, http://www.non-lethal-weapons.com/sy05index.html.

8: The Search for the Magic Bullet

1 I had heard the term "magic bullet" a few years before using it in a presentation for the 1st European Symposium on Non-Lethal Weapons, September 25–26, 2001 in Ettlingen, Germany. A narrative version is available in hardcopy at: Charles "Sid: Heal, "What Will the 'Magic Bullet' Look Like?" *Non-Lethal Weapons: New Options Facing the Future*, Fraunhofer-Institut für Chemische Technologie (ICT). Ettlingen, Germany (September 2001), 15-1.

2 Taken from Aesop's fables, this metaphor is based on a story of mice deciding to put a bell on a predatory cat to provide a warning when it approached. Although sound in concept, the application is full of danger, since there is great risk to the

one assigned to attach the bell. The parable succinctly identifies the problems with being first in any risky undertaking.

3 David Klinger is currently Associate Professor, Department of Criminology and Criminal Justice, University of Missouri-St. Louis.

4 This situation is far more prevalent in law enforcement circles than in the military.

5 David Klinger, PhD. "Revenge Effects and Less-Lethal Munitions," *Command* (1998): 6–9.

6 This also includes nonlethal devices. Over the years, I have been involved in countless discussions with developers and manufacturers of nonlethal weapons, who have a usable device but are unable to attract customers. Eventually, I developed a list of all the currently available force options by what it cost to employ them for one "dose," so that comparisons became more obvious. This included the cost of the launcher and munitions, as well as the number of "applications" required to be effective. Using the same criteria for the new device, it became conspicuous as to which nonlethal weapons it was competing with. For one example, a dazzling laser was being offered at well over $3,500. Although repeated applications were free, the cost of the "launcher" was still prohibitively expensive when compared with a TASER®, which cost more than $800 and an additional $30 per dose (cartridge). Thus, the only strong selling point was that a device provided some advantage over the competition that justified the extra expense. This was often a showstopper.

7 I was introduced to this rule as a young rookie in South Central Los Angeles in 1980, but it has been around since at least the mid-1970s. I have heard it called the "Tueller Drill" after Salt Lake City, Utah police instructor Dennis Tueller, who did extensive research to verify it. It has withstood the test of time, the scrutiny of science, and the judgment of courts. For more information on this factor, see "Is the 21-Foot Rule Still Valid When Dealing With an Edged Weapons? Part 1," *Force Science News #17*, April 22, 2005 https://www.forcescience.org/2005/04/is-the-21-foot-rule-still-valid-when-dealing-with-an-edged-weapon-part-1; Part II of Special Edged Weapons/21-Ft. Rule Series, *Force Science News #18*, April 29, 2005 https://www.forcescience.org/2005/04/is-the-21-foot-rule-still-valid-when-dealing-with-an-edged-weapon-part-2/; and Charlie Mesloh, Mark Henych, and Ross Wolf, "Less Lethal Weapon Effectiveness, Use of Force, and Suspect & Officer Injuries: A Five-Year Analysis," A Report to the National Institute of Justice, Florida Gulf Coast University, Weapons & Equipment Research Institute (September 2008), 23. Although there are also a number of court cases, one of the more recent that deals with the danger related to distance is: Estate of Larsen v. Murr, 511 F. 3d 1255 (10th Cir. 2008) https://www.courtlistener.com/opinion/170135/estate-of-larsen-ex-rel-sturdivan-v-murr/.

8 Admittedly, this is somewhat arbitrary. In this case, severe injury was defined as those that might prove permanently disabling or life-threatening. The weight was set at 530 grams, or about 1.2 pounds, or for us street cops about the weight of one-third of a brick. It needs to be noted, however, that smaller objects are capable of being thrown as much as a hundred yards. Because Los Angeles is nearly all paved, rioters would bring their own missiles, such as spark plugs, wheel weights, and golf balls, to throw at us. (They are cheap and can be purchased in

bulk.) These may not kill, but you can still end up with some pretty nasty cuts and bruises.

9 This is a general guideline, since the weight of the object is only one of the factors involved. Where it hits the body and how fast it is moving are also important. The guideline was determined by a study done jointly by Penn State's Applied Research Laboratory and the Los Angeles Sheriff's Department in the fall of 2000. For more information see: Dr. John M. Kenny, *et al.*, "The Attribute-Based Evaluation (ABE) of Less-Than-Lethal, Extended-Range, Impact Munitions," Pennsylvania State University, Applied Research Laboratory and Los Angeles Sheriff's Department, State College, PA (February 15, 2001), 16, https://www.scribd.com/document/59570810/Abe-Report.

10 Testimony before the joint hearing of the Crime Subcommittee of the House Judiciary Committee and the National Security International Affairs and Criminal Justice Subcommittee of the House Government Reform and Oversight Committee, August 1, 1995.

9: Classifying Nonlethal Options

1 The International Law Enforcement Forum (ILEF) was initiated by Penn State's Institute for Non-Lethal Defense Technologies (INLDT) in 2001 with assistance from the Los Angeles Sheriff's Department and the UK's Police Scientific Development Branch and Home Office. Using the Internet and subject matter experts from a number of countries seeking to establish common ground in developing and employing nonlethal options, an Electronic Operational Requirements Group (EORG) has been attempting to develop international standards for development, testing, training, and operational use.

2 One issue that was never resolved was that in an ideal taxonomic system each option would have one, and only one, assignment. Because of the versatility of the various options, this issue has never been completely resolved, nor do I have an answer for it.

3 For a thorough description of the differences between lethal and nonlethal force see Chapter 1, "Nonlethal Confusion and Controversies."

4 Of note is that information systems have been intentionally excluded. Whereas information warfare can be considered a type of nonlethal alternative when it seeks to delay, deny, manipulate, or corrupt information without actually harming people, it is largely held that there are more differences than commonalities and so should be considered separately from other nonlethal options. One of the more cogent arguments for this position can be found in Colonel Joseph Siniscalchi's thesis for Air War College, "Non-Lethal Technologies: Implications for Military Strategy." See http://www.fas.org/man/dod-101/sys/land/docs/occppr03.htm.

5 Of interest is that, using this methodology, substances like sticky foam would be categorized as mechanical devices. Sticky foam is an expansive chemical that works as a human adhesive. Consequently, the predominant method of operation is not pharmaceutical but in restraining motion, and is more appropriately placed in the mechanical category than in the chemical category.

6 Alternative suggestions have included "living systems," "live systems," and

"insect-animal based systems." For unknown reasons, the term "vivi-systems" is controversial.

7 Canines have not only proven to be an exceptionally effective nonlethal tool, but accidental deaths are extremely rare. In fact, in the history of US law enforcement only one accidental death from a law enforcement canine has ever been recorded.

8 Like all categories, the criterion that distinguishes the various categories is the predominant method of operation. Consequently, the use of pheromones to attract noxious bugs and the like are chemicals, but the predominant method of operation is by using a living organism, hence the assignment to the vivi-system category and not the chemical category.

9 "Non-stabilized" are sometimes called "slicks" and refer to those projectiles with no method of in-flight stabilization. The pad-type stun bags are a good example. Spin-stabilized projectiles are those that spin in flight to establish and maintain ballistic stability. Some use fins whereas others are fired from a **rifled barrel**. Fin-stabilized projectiles are made from a stiff material, such as hard rubber or plastic, and have fins that provide aerodynamic stabilization. Drag-stabilized projectiles are those that have a "tail," much like a kite and are intended to maintain some in-flight stabilization. Ring airfoils are projectiles that resemble an aircraft wing folded into a donut-shape and provide in-flight lift and stabilization.

10: Nonlethal Impact Weapons

1 They are frequently referred to by other names as well, including specialty impact munitions, blunt impact munitions, kinetic energy impact projectiles, and extended range impact munitions. In choosing a term for this book I opted for the simplest description that included impact, since so many other options involve movement that the distinguishing characteristics would be obscured.

2 A variation of this group uses leather to encase lead shot, metal, or wood, usually with a flexible handle. With names like saps, blackjacks, or bludgeons, they reduce the risk of lacerations, while still delivering a serious blow. Few contemporary law enforcement agencies still allow their use as a nonlethal force option.

3 As a matter of fact, the most modern handheld impact weapons can still be lethal. It is only the skill and intent of the person who is wielding them that prevent permanent injury or death.

4 Truncheons are even mentioned in Shakespeare's *Hamlet.*

5 Some cite whips as the earliest form of nonlethal weapons. However, I have omitted them because they fall short of the modern definition, since they commonly leave permanent scars. Notwithstanding, they have been used throughout history and are recorded in the Book of Isaiah, which dates back to the eighth century B.C.E. Indeed, it is recorded in Matthew 2:15 that Jesus made and used a whip to drive the moneychangers from the temple area in Jerusalem.

6 The sjambok is the traditional leather whip of South Africa (although it is common nowadays for the sjambok to be made of plastic). It is approximately three to five feet (1–1.5m) long and one inch thick (25mm) and is reviled as a symbol of Apartheid. The bullwhip is made of braided leather and was originally a farmer's tool for herding livestock. Bullwhips are typically ten feet (3m) and even longer. The cat-o'-nine-tails is a multi-tailed whip with a rigid

handle. Each of the tails can be knotted or have objects laced into them to inflict more injury. Cat-o'-nine-tails whips can be traced to at least the late Renaissance period, although similar whips without the name can be traced to the Roman times. Whipping is still used as a form of punishment in a number of countries, especially Southeast Asia and the Caribbean. Commonly called caning or birching, the most well known incident of this action occurred in 1994 in Singapore when Michael Peter Fay, an eighteen-year-old American, was sentenced to six strokes of the cane after pleading guilty to vandalism. The sentence was later reduced to four strokes.

7 Those nonlethal impact munitions that require a separate launcher have not been as popular as those using shotguns or 37/40mm launchers. Law enforcement does not like them as well because the launchers tend to be expensive and fire only projectiles from the manufacturer. The so-called one-trick pony requires resupply from a single source. The military community favors "dual-use platforms": that is, weapons platforms that can deliver both lethal and nonlethal munitions, such as the M-203 5.56mm and 40mm. Consequently, systems that provide only a nonlethal capability are not appealing.

8 These knee knockers are one of the earliest versions of nonlethal impact munitions and were reportedly first used in Hong Kong by the British in 1958.

9 So-called sponge grenades are simply soft-nosed projectiles fired through a 40mm launcher. They are intended for direct fire and are highly accurate, with a longer range in comparison to other nonlethal weapons, and they are one of the most consistent in energy transfer.

10 Determined rioters and provocateurs quickly learn that the effects from stingballs, even in close proximity, are relatively mild and can be countered by simply wearing heavy clothing and turning away. One of the most effective deterrents is to exploit the countermeasures with other options, including more than one stingball or flashbang. Furthermore, one of the more effective techniques developed by law enforcement is to use stingballs to separate a provocateur from other rioters and then arrest him before he has an ability to recover. The expression, "Deploy in volleys, exploit with tactics," has been a successful principle for effectively using flashbangs and stingballs. Even so, they are indiscriminate options that strike everyone in the vicinity with no distinction between combatants and bystanders.

11 Because projected energy is virtually synonymous with directed energy, this group might be logically placed under the category of directed energy. Because the PEP works with impact, however, I placed it under impact weapons. The rationale is that the predominant method of operation is with impact not radiation. In this case, radiation is how it is launched, not how it works.

12 Plasma is called the fourth state of matter in that it is neither liquid, solid, nor gas. In my limited understanding, it is a highly ionized gaseous state in which electrons are not bound to an atom or molecule, consequently it is highly conductive. This creates all kinds of possibilities with pulsed energy being only one.

13 A similar technology, called the active denial technology, was unveiled by the Pentagon in April 2001, but as of this writing it is still not being employed. Active denial devices are similar to the PEP in that they use directed energy, but actually

work by radiating electromagnetic energy in stimulating nerve endings to create a sensation of pain.

14 One best practice used in custodial and correctional institutions (and other fixed positions) is the use of range markers. These can be as simple as chalk or painted numbers or as sophisticated as colored signs and icons.

15 This phenomenon was first noted by Steve Ijames, one of the most knowledgeable experts on the effects of nonlethal force option in the world. Although separated by thousands of miles, Steve and I have exchanged information and collaborated on nonlethal force options for decades as we rose through the ranks in our individual departments. Steve remains one of the most insightful experts in the use of force in general and nonlethal force in particular.

16 It is estimated that about 10 percent of these wounds result in death because of damage to the central nervous system, while 50 percent result from bleeding to death. The general rule is that these injured are "treated in the emergency room but saved in the field." Hence, anything that can be done to reduce the urgency or expedite the medical treatment will save lives. Besides the altruistic reasons for saving the life of a combatant, medical evacuations are hugely labor- and time-intensive and will compete with tactical objectives.

17 Since at least 1898, scientists and developers have attempted to identify a single energy threshold that could be used to separate minor injuries from more serious ones. Various numbers have been suggested. For instance, in the 1970s the US Army Land Warfare Laboratory reported that 30–90 foot-pounds is "dangerous" and anything above 90 foot-pounds would cause "severe damage," but 58 foot-pounds has also been used. In practice, the amount of energy alone is not enough to prevent serious injuries. Suspects routinely absorb multiple impacts with projectiles dumping well over 90 foot-pounds of energy and suffer only bruising, even though a young college student (Victoria Snelgrove) was accidentally killed in 2004 with a projectile that had an impact of only about 25 foot-pounds. For more information see: Dr. John M. Kenny, *et al.*, "Non-Lethal Weapon Injury Characterization," Human Effects Advisory Panel, Institute for Non-Lethal Defense Technologies, Applied Research Laboratory, Penn State, State College, PA (August 30, 2004), 4 and 19–20.

18 Of note is that whereas both the 37 and 40mm dimensions are approximately 1½ inches, the 37mm has been the caliber most used by American law enforcement; but 40mm is the NATO standard and used by the military community. The most likely reason is that launchers used by the military prior to World War II were 37mm and became military surplus when America switched to the 40mm NATO standard. Consequently, law enforcement inherited huge numbers of free, or nearly free, 37mm launchers when they were no longer of value to the military. Also, early interest in nonlethal munitions was nearly entirely focused on the law enforcement community. Understandably, developers and manufacturers designed munitions to fit the market. With the increasing interest of the military community in similar munitions, coupled with the need for replacing aging law enforcement launchers, the 40mm is gradually becoming the standard for both law enforcement and the military.

19 This is also why the new expandable and lighter handheld batons can hit harder than the heavier wooden ones. For more information, see Charlie Mesloh, Mark

Henych, and Ross Wolf, "Less Lethal Weapon Effectiveness, Use of Force, and Suspect & Officer Injuries: A Five-Year Analysis," A report to the National Institute of Justice, Florida Gulf Coast University, Weapons & Equipment Research Institute (September 2008), 27.

20 For just about everyone else in the world, it is measured in joules: one foot-pound equals 1.355817948 joules. Thus, 25 foot-pounds of energy would be about 34 joules and 90 foot-pounds would be about 122 joules.

21 One good example is that a .22 caliber bullet releases about the same amount of energy as a 12-gauge stun bag. As can be imagined, however, the injury to a human body will differ significantly. For more information, see David K. Dubay, Director of Research, Defense Technology/Federal Laboratories, and Paul J. Marquard, Department Chair, Physics, Casper College, "Kinetic and Impact Parameters of Less-Lethal Munitions" (Caspar, WY: Defense Technologies 2003 Specification Manual), 3.

22 Glover, Scott, and Matt Lait, "Police Clash May Cost Woman an Eye," *Los Angeles Times*, Sec. B, Col. 1, December 1, 2000.

23 Kenny, Dr. John, M., *et al.*, "The Attribute-Based Evaluation (ABE) of Less-Than-Lethal, Extended-Range, Impact Munitions," Pennsylvania State University, Applied Research Laboratory and Los Angeles Sheriff's Department, State College, PA, February 15, 2001.

24 A durometer is a measure of hardness of some material returned with a value from 0 to 100, with the higher the number, the harder the material. It takes its name from the instrument used for measuring it and measures the indentation in the material when compressed. Unfortunately, that's about as simple as it gets, because there is more than one scale. At the risk of oversimplification, the "A" scale is for soft rubber and plastics and the other scales are used for harder materials.

11: Chemical Agents

1 Adrienne Mayor, *Greek Fire, Poison Arrows & Scorpion Bombs* (New York: Overlook Press, 2003), 225.

2 It was also commonly used by Japanese swordsmen to disable an opponent and make it easier to kill him. Using the modern definition, which requires that a nonlethal weapon be explicitly designed and primarily employed so as to minimize fatalities and permanent injury, uses like these fail to meet the criteria of a nonlethal weapon.

3 Tear gas has become a generic term to include nearly all riot control agents. In reality, many of the modern compounds work in other ways. The chemical used to capture the Parisian bank robbers was ethyl bromacetate, but it has been estimated that about thirty riot control agents had been developed at the time. Paris police also used them to dispel rioters before World War I. Tear gas had another nonlethal advantage in training soldiers to don and clear gas masks, since it provided an irrefutable, but nonlethal, proof of failure. It is still used for the same purpose by both law enforcement and the military today.

4 Eugene Nielsen, "The Selection and Tactical Employment of Less-Lethal Chemical Munitions: A Basic Overview," *Police and Security News* 13, no. 6 (November/December 1997): 41–46.

5 For the sake of simplicity, I have included soporific and hypnotic chemicals in this group, as they are simply stronger versions of calmatives. There have been some discussions on whether obscurants, such as smoke, should be a considered another group of the nonlethal chemical agents category. Because obscurants are not employed with the intent of causing a pharmaceutical interaction with the body, I have excluded them. Using the same criteria, other chemicals, such as filter-cloggers (engine stoppers) or super-lubricants, are more appropriately applied in the mechanical category, since they work by inhibiting or restraining motion.

6 Although pain is a natural consequence and enhances the debilitating effects, it is more of a byproduct than the intended result. Chemical agents are one of the few nonlethal options that do not rely on pain for effectiveness.

7 More precisely, chloroacetophenone (CN) is a lachrymatory agent, or just lachrymator. The term comes from the Latin *lacrima*, meaning "tear," hence the term *tear gas*. CN was used almost exclusively for more than forty years, which by that time the term "tear gas" and CN had become nearly synonymous. Since CS so closely mimicked the symptoms of CN, there was no need to change the terminology as it became more popular and, even though OC is usually identified as an inflammatory, it has been indiscriminately included in the generic grouping of "tear gas."

8 The US Army officially adopted CS for combat training and riot control purposes in 1959. See "Flame, Riot Control Agent, and Herbicide Operations," FM 3-11.11, MCRP 3-3.7.2, Headquarters, Department of the Army and United States Marine Corps, Washington DC (March 10, 2003), 6-1.

9 Without going into excruciating detail, CN has a lethal concentration threshold of 14,000 milligrams per cubic meter. CS has a lethal concentration threshold of 25,000 milligrams per cubic meter, making it nearly twice as safe as CN. The lethal concentration threshold is usually referred to as LCT, or LC50, which means that the lethal concentration is expected to kill 50 percent of those exposed for a one-hour duration. Likewise, the incapacitation threshold, usually referred to as ICT, is also identified as ICT50. The ICT for CN is 20 milligrams per cubic meter while the ICT for CS is 10 milligrams per cubic meter, meaning that CS requires only half the concentration to affect 50 percent of the exposed population when compared with CN. In the simplest terms, it takes about half the amount of CS than CN to have the desired effect but more than twice as much to be lethal.

10 A fair estimation is that more than 95 percent of American law enforcement agencies are authorized to use pepper spray, which equals or exceeds those authorized to use batons.

11 Although unusual, complete decontamination times for OC can extend for an hour or longer. This is also why American law enforcement usually prefers CS over OC for riot control, since decontamination requires a separate logistical component and creates a competing mission with tactical objectives.

12 One other advantage is that there is no known lethal concentration. Although I'm sure some enterprising critic will argue that even water has a lethal concentration, as when you are drowning, no one knows for sure what that is with OC.

13 See "Flame, Riot Control Agent, and Herbicide Operations," FM 3-11.11,

MCRP 3-3.7.2, Headquarters, Department of the Army and United States Marine Corps, Washington DC (March 10, 2003), 6–2.

14 Various sources also attribute the synthesis of DM to German chemist, Heinrich Otto Wieland, in 1915 as well as Roger Adams in 1918. For more information, see Neil Davison, *The Early History of "Non-Lethal" Weapons*, Occasional Paper No. 1, Bradford Non-Lethal Weapons Research Project (BNLWRP), Department of Peace Studies, University of Bradford, Bradford, U.K (December 2006), 9; and "The Evolution and Development of Police Technology," A Technical Report prepared for the National Committee on Criminal Justice Technology, National Institute of Justice, Washington DC (July 1, 1998), 39.

15 Somewhat ironically, the protestors were unemployed American World War I veterans who had gathered in Washington DC to demand political changes, and the users were also US Army soldiers.

16 Malodorant programs in the US Department of Defense can be reliably dated back to the 1960s. It has only been recently, however, that the need for them as a nonlethal option has generated a new impetus. Consequently, private developers have begun providing commercial malodorants for a variety of functions, such as keeping animals away from gardens or gang members and transients out of vacant buildings. Commercial malodorants are available as creams or liquids and must be applied by hand.

17 For those who don't know, Typhoid Mary was a cook who became famous, or more accurately infamous, for spreading highly contagious typhoid fever. Even though she was a carrier of the deadly disease, she refused to take suitable precautions and remained healthy while infecting scores of people. Her name became famous as a generic term for the carrier of any dangerous disease who continues spreading it while refusing appropriate safeguards.

18 A new concept called a taggant can also be included with a chemical like a malodorant. A taggant is simply a nonreactive substance that can be traced. Some taggants glow, like a chem-light, whereas others are invisible and can only be seen with infrared or ultraviolet lights. Infrared allows suspects to be tracked through night-vision equipment, without their being aware, and arrested at a more convenient time and location. Still other taggants provide a visual ability to trace a single projectile to the person it hit. Nicknamed portable DNA, this capability is critical for the law enforcement community, who must be able to testify after an incident that this particular person, and no other, did a specific act for which he should be prosecuted and this is how they are positive it was him. This is called the continuous chain, and is required for juries to convict.

19 Mercaptan is added to natural gas, which is naturally odorless but deadly, as an olfactory signal to alert you of the danger. Accordingly, it has also been suggested that a malodorant be developed for other deadly but odorless plume clouds, especially for acts of terrorism. Because many deadly plume clouds are invisible, odorless, and tasteless, detection and/or evacuation is difficult—extraordinarily so when the cloud is dispersed through heating and air conditioning ducts in large buildings. In a malodorant being released into the same area, the smell provides both a danger signal and "self-evacuating" assistance, while superimposing a harmless stench over an otherwise undetectable but deadly plume.

20 Malodorants were used extensively as a riot control agent by the Israelis in August

of 2008. Called "skunk" by the Israelis, the term provides a strong clue as to how badly it smells. It proved so successful that it is being distributed to local police arsenals. See, Kobi Ben Shimon, "Making a Stink," Haaretz.com, September 4, 2008, https://www.haaretz.com/1.5021400. See also, David Hambling, "Israel Unleashes First 'Skunk Bomb,'" *Wired Magazine*, September 21, 2008, https://www.wired.com/2008/09/skunk-attack.

21 Although the actual agent has never been completely revealed, it is believed that it was a derivative of fentanyl, a short-acting but extremely powerful and commonly used anesthetic. Nor were emergency medical responders aware of the specific drug used, even though a counteragent, naloxone, would likely have reduced casualties had it been immediately administered. See Susan B. Glasser and Peter Baker, "Russia Confirms Suspicions About Gas Used in Raid; Potent Anesthetic Pumped into Theater; 2 More Hostages Die from Drug's Effects," *The Washington Post*, October 31, 2002, A15.

22 (Then) Lt. Richard "Odie" Odenthal, Los Angeles Sheriff's Department, Emergency Operations Bureau.

23 For a more thorough discussion of this subject, see, Dr. Joan M. Kaloski, Dr. W. Bosseau Murray, and Dr. John M. Kenny, "The Advantages and Limitations of Calmatives for Use as a Non-Lethal Technique," College of Medicine, Applied Research Laboratory, Pennsylvania State University, State College, PA, October 3, 2000 9.

24 As a matter of fact, had safe and effective calmative agents been available, they may have been used during the assault on the Branch Davidian compound in Waco, Texas in 1993, nearly a decade before the Dubrovka Theater incident in Moscow. As reported in *Time* magazine, "[Janet] Reno [US Attorney General at the time] continued to press about the dangers of exposing people to gas. Anesthetic gases might knock people out, but there was no guarantee that they would wake up, ever, especially the small children. Strong men would be knocked out last, or not at all. The FBI brought in a leading specialist on the toxicology of tear gas, whom Reno debriefed for hours. She approved the use of tear gas only after being assured that the form the FBI was using was not permanently harmful, carcinogenic or a possible cause of birth defects." Quoted in Nancy Gibbs, "Oh, My God, They're Killing Themselves: Waco Comes to an End," *Time*, May 3, 1993, http://content.time.com/time/magazine/article/0,9171,978360,00.html.

12: Mechanical Options

1 Of interest is that using the predominant means of effectiveness as the distinguishing characteristic, these chemical substances function by restraining the intended motion of something rather than employing a pharmaceutical reaction, and so are more appropriately placed in the mechanical category rather than the chemical one.

2 PCP (Phencyclidine hydrochloride) was invented and first used as an anesthetic agent in the early 1950s and remained available as an animal tranquilizer until 1978. An estimated 25mg per kilogram of body weight allowed general surgery in forty-five minutes while a patient remained conscious but without experiencing pain. Unfortunately, the patients also experienced hallucinogenic and dissociative effects. PCP began appearing on the streets as a recreational

drug in the mid-1970s and was most commonly smoked, as this method allowed a user to control the amount of drug needed to get high without overdosing. Conventional weaponless control techniques rely heavily on pain compliance but were nearly completely ineffective against even small-bodied combatants, who felt no sensation of pain. Virtually overnight, PCP rendered every nonlethal option to that time ineffectual, often resulting in serious injuries to police officers and suspects alike, as well as promoting the need for lethal force. PCP is often cited as one of the driving motivations for developing effective nonlethal alternatives by American law enforcement.

3 On at least two occasions, I have personally witnessed unprotected role players suffer head wounds serious enough to require stitches when struck by these weights.

13: Electrical Options

1 Like the term "rubber bullets" the term "stun gun," is media-inspired. It has captured the imagination of the public but is neither scientific nor wholly descriptive. The term is used to refer to any nonlethal electrical force option that employs launched projectiles, regardless of whether or not it resembles a gun.

2 The term "TASER" is capitalized because it is an acronym coined by the inventor, Jim Cover, which he named after a childhood hero. Written at the turn of the twentieth century, the term "Thomas Swift's Electric Rifle" is the title of one of the more than hundred volumes written by the Stratemeyer Syndicate. (The "A" was added to make a pronounceable acronym.)

3 These figures exceeded any nonlethal force option in our arsenal and, to my knowledge, have never been surpassed. Moreover, pepper spray, our next best option, affected the subjects for as long as forty-five minutes, not to mention contaminating our uniforms and radio cars while arresting and transporting. Whereas the TASER® remains a relatively short-ranged weapon (the tethered darts can travel about 25 feet), our typical engagement ranges have been 12–18 feet, making it an excellent option inside buildings and when dealing with combative subjects.

4 Having spent an entire career in law enforcement, with my most formative years on the streets of South-Central Los Angeles, I feel qualified to extol the virtues of the TASER® as a nonlethal option for domestic law enforcement applications. Indeed, the importance of TASERs® in providing effective nonlethal force alternatives for the law enforcement community can hardly be overstated. Although this technology still falls well short of the "magic bullet" described in Chapter 8, it is unsurpassed in safety and effectiveness within the effective engagement ranges and against the fiercest combatants. Those who demand moratoriums and restrictions condemn us to employing far more primitive and harsher alternatives. The problem remains, only the means to resolve it has changed. The standard is not perfection; it is the alternative.

14: Biological Options

1 Charlie Mesloh, "Barks or Bites? The Impact of Training on Police Canine Force Outcomes," *Police Practices and Research: An International Journal* 7, no. 4 (2007): 323–325, https://www.tandfonline.com/doi/

abs/10.1080/15614260600919670. Note: This is a conservative estimate in that some modern DNA studies indicate that the transformation between wolves and dogs may have occurred as long as 130,000 years ago.

2 William F. Handy, *et al.*, "The K-9 Corps: The Use of Dogs in Police Work," *Journal of Criminal Law and Criminology* 52, no. 3 (Fall 1961): 328–337.

3 Charlie Mesloh, phone Interview, May 22, 2019. Note: This number is based upon the research and knowledge available. There are no reporting requirements and no organizations that track the number of police service dogs on a national level.

4 The term "battlespace" is used by the military community to describe the domain or realm where an adversary can be acquired and engaged. Although it may be an emotion-arousing term for some, it is clearly descriptive for any environment in which an adversary may be encountered, including law enforcement applications.

5 Daniela Barberi, Jennifer C. Gibbs, and Jennifer L. Schalley, "K9s Killed in the Line of Duty," *Contemporary Justice Review* 22, no. 1 (2019): 86–100.

6 This estimate includes all canine line of duty deaths, including accidental.

7. Vince Guerrieri, "Dog Cleared of Profiling, Back on Duty," *Pittsburgh Tribune*, August 17, 2002.

8 Deputy Romeo Ingresso, Los Angeles Sheriff's Department.

9 Kerr v. City of West Palm Beach, 875 F. 2d 1546, 11th Cir., 1989.

10 *Ibid.*

15: Directed Energy

1 Although it is possible to reflect a directed energy wave, such as using mirrors with light, as yet, no practical method has been developed for enhancing nonlethal force options.

2 It must be noted that this description is overly simplified. The human eye is extremely complex and is affected not only by light, but age, injury, illness, drugs, and even emotional arousal. This brief description is simply to emphasize the difficulties involved in developing safe and effective nonlethal force options.

3 Epilepsy is a disorder of the central nervous system that may result in seizures. Although photosensitivity is higher in people diagnosed with epilepsy, especially in people forty years old and younger, it remains rare in the population at large. The ill-effects related to an existing medical condition are 3rd Order Effects and are impossible to foresee in individuals, especially those engaged in activities that require forceful interventions. (See "Orders of Effects" in Chapter 5.) For a detailed understanding, see, A. Martins da Silva and Barbara Leal, "Photosensitivity and Epilepsy: Current Concepts and Perspectives—A Narrative View," *Seizure: European Journal of Epilepsy* 50 (August 2017) 209–218, https://www.seizure-journal.com/article/S1059-1311(17)30252-2/fulltext.

4 Wavelength is closely related to frequency and changing one changes the other. For simplicity and clarity, just understand that shorter wavelengths will have the highest frequencies and vice versa. Because light travels at a given speed (186,282 miles per second), a different light wavelength *must* have a different frequency. Thus, colors of light can be identified in distance (nanometers) or frequency (hertz).

5 Similar to LEDs, lasers can be built in different configurations but with far more difficulties in making them eye-safe. All lights have the potential of causing

serious injury subject to the aperture, exposure, and intensity. Currently, lasers cost as much as ten times their LED equivalent and that is unlikely to change much in the foreseeable future.

6 As a point of interest, the decibel scale is logarithmic, meaning that the scale increases (or decreases) in powers of ten. Therefore, an increase of 10 dB is a tenfold increase in loudness. Accordingly, an increase of 6 dB is twice as loud even though it will take about 10 dB for humans to perceive the sound as twice as loud. This is especially important in selecting and deploying nonlethal force sound options, since 100 dB at fifty feet will not be 50 dB at a hundred feet, but rather 94 db.

7 There are many other ways of using sound as a nonlethal force option, such as using modulation (varying the frequency, volume, etc.). As with light, the intensity of the sound and duration of the exposure are major factors.

8 Studies have revealed that these frequencies are typically in the range of 2000 to 5000 Hz, which not surprisingly, is also where human ears are especially sensitive.

9 To include the United States, Canada, Australia, and a number of European countries.

10 Currently, the recent injuries suffered by US diplomats in China and Cuba are suspected to be a result of ultrasound being used as a listening device. Although maybe not intentional, the dangers are demonstrably apparent.

11 Dr. John M. Kenny, *et al.*, "A Narrative Summary and Independent Assessment of the Active Denial System," Penn State, Applied Research Laboratory, State College: Human Effects Advisory Panel, 2008.

12 Active Denial System FAQs, US Department of Defense, Non-Lethal Weapons Program https://jnlwp.defense.gov/About/Frequently-Asked-Questions/Active-Denial-System-FAQs/. It is believed that the RF wave may reflect off the forearms and biceps of individuals like a parabolic antenna and exacerbate the effects. Regardless, both subjects fully recovered without anything more serious than some painful but temporary blisters.

13 This is a valid concern and is also commonly directed at TASERs®. It is only human nature to seek the easiest method. Notwithstanding, that is a policy, training, and supervision problem, not a technological one.

14 Patrick Tucker, "The Pentagon is Making a Ray Gun to Stop Truck Attacks," *Defense One*, April 24, 2018, https://www.defenseone.com/technology/2018/04/pentagon-making-ray-gun-stop-truck-attacks/147702/.

16: Conclusions

1 A demonstrable desire decisively refutes claims of indifference.

Index

Note: The Glossary of Terms is not included in this index. Footnote references are referenced by "fn" followed by the letter citation (i.e., 35fna is the footnote citation "a" on page 35). Endnote references are indicated by "n" followed by the endnote citation (i.e., 149n4 is endnote number 4 on page 149). Pages with more than one endnote with the same citation number are indicated with a letter (i.e.,161n3a refers to the first endnote citation 3 on page 161).

About the Author

Charles "Sid" Heal is a retired Commander from the Los Angeles Sheriff's Department with nearly 33 years of service in law enforcement, nearly half of which has been spent in units charged with handling law enforcement special and emergency operations. Sid has earned three college degrees and is a graduate of the California Peace Officer's Standards and Training, Center for Leadership Development, Command College, and the FBI National Academy. He is the author of *Sound Doctrine: A Tactical Primer*, *Field Command*, and *An Illustrated Guide to Tactical Diagramming*, as well as more than 120 articles on law enforcement issues. He has appeared on numerous television newscasts and documentaries, and been quoted in many periodicals and newspapers. Additionally, he has been a featured speaker at numerous conferences in Canada, Germany, England, Scotland, Ireland, Israel, Brazil and Argentina, as well as throughout the United States.

About the Publisher

LANTERN PUBLISHING & MEDIA was founded in 2020 to follow and expand on the legacy of Lantern Books—a publishing company started in 1999 on the principles of living with a greater depth and commitment to the preservation of the natural world. Like its predecessor, Lantern Publishing & Media produces books on animal advocacy, veganism, religion, social justice, psychology and family therapy. Lantern is dedicated to printing in the United States on recycled paper and saving resources in our day-to-day operations. Our titles are also available as ebooks and audiobooks.

To catch up on Lantern's publishing program, visit us at www.lanternpm.org.

facebook.com/lanternpm
twitter.com/lanternpm
instagram.com/lanternpm